THE
GAMe

Dennis —
Thanks for all
you do to help our Vets!

Robert D. Shindley

THE
GAMe

UNRAVELING
A MILITARY
SEX SCANDAL

ROBERT D. SHADLEY
MAJOR GENERAL, U.S. ARMY (RETIRED)

ISBN 13: 978-1-59298-996-6

Library of Congress Catalog Number: 2013902843

Printed in the United States of America

First Printing: 2013

17 16 15 14 13   5 4 3 2 1

BEAVER'S
POND
PRESS

Beaver's Pond Press, Inc.
7108 Ohms Lane
Edina, MN 55439–2129
(952) 829-8818
BeaversPondPress.com

To order, visit: BeaversPondBooks.com
or call (800) 901-3480. Reseller discounts available.

Jimmy D. Ross, General, United States Army
May 23, 1936, to May 2, 2012
Boss, mentor, and friend
"Bob, you've got to tell this story."

Elinor (Ellie) K. Shadley
Wife, mentor, friend, and the best thing that ever happened to me
"Bob, the first thing you have to do is take care of those women."

CONTENTS

ACKNOWLEDGMENTS

I had been thinking about writing this book for years. We would often fantasize at our crisis action team meetings about writing a book and who would play which parts in the movie. Humor got us through a tough, nonhumorous situation. I began writing in February 2011, and the first version was almost 1,000 pages. I was told a book with that many pages would not fly in the marketplace. So, I began the process, with much help from several great experts, to pare the manuscript down. What follows is an accurate accounting of what was and still is known as the Aberdeen Sex Scandal. I used my personal notes and unclassified documents to provide the basis for this chronicle. I submitted this manuscript to the Department of the Army for security review, and they confirmed "no classified information was identified."

Over the years, I have learned it is impossible to thank everyone who assists you in a particular endeavor or in life in general. Invariably, someone is left out, and I feel terribly that they might think they were not important. In that regard, I would just like to say a heartfelt thanks to a few groups of people:

The vast majority of the great military and civilian employees and their families at the US Army Ordnance Center and School, who never did anything wrong, worked hard every day to do the right things, and continue to do so today.

Our crisis action team, who provided sound and caring advice and kept us on the right track of working our objectives: Take care of the victims, let the victims identify the alleged perpetrators and let the judicial system work its due process, and identify corrective actions to preclude recurrence.

Our spouses and significant others, who helped 24/7 by loving us and supporting our families while we worked long hours to do the right thing.

Friends, colleagues, subordinates, contemporaries, and bosses who supported me, as well as those who did not support me for reasons they thought were right in their own minds. All made me a better person.

The people who motivated me to complete this project and provided guidance on how to maneuver the publishing process.

And to the women in our military—things will get better. They can and they will if we all work to prevent sexual assaults.

INTRODUCTION

The drill sergeants met the buses as they rolled in, callously scouting which female members of the new class of trainees they would try to have sex with. It was so organized and endemic a practice, they had given it a name: Playing the Game. It was also known as GAM—Game à la Military.

For the female victims, it was not a game. They were the victims of an abuse of power by men who were supposed to be their immediate leaders in their chain of command. Instead of leaders, they found lechers.

GAM was the name for a deeply embedded system of sexual harassment and assault going on in the Army for many years. When confronted with this knowledge in my new command at Aberdeen Proving Ground in Maryland in 1996, I was determined to protect the victims, allow the investigative and judicial systems to identify the perpetrators, and prevent further abuses in the very uneven power balance between drill sergeants and young trainees.

I never imagined my efforts to correct these serious abuses would expose me to criticism and reprimand from the very Army I had loved and served for more than 30 years. I would be labeled a racist by some organizations and vilified by others in the press. My actions and those of my team were scrutinized by the media, private organizations, members of Congress, the Office of the Secretary of the Army, and the Office of the Secretary of Defense. Some had praise. Others had criticism. What follows is a true, carefully documented account of the sexual misconduct scandal at Aberdeen Proving Ground and beyond.

My years in service in the United States and abroad included a tour

1

in Vietnam as well as in Southwest Asia in support of Desert Shield and Desert Storm. The work was difficult and demanding. However, the 11 months I spent at Aberdeen immersed in the sex scandal were my most stressful in the military. The consequences and results of my efforts, and those of our team, were not what we had hoped and worked for. Even so, I would not change my decision to seek justice for the victims of abuse.

I was disappointed in my Army. It appeared our senior leadership was more concerned with making the problem go away and protecting the institution than with solving the problem and protecting our soldiers. Tragically, we are now living with the consequences, as well documented in the media. Sex scandals in the military are a recurring event.

Using the Department of Defense estimate as reported in the media, 18,000–19,000 service members in the military each year are subjected to sexual assaults (felonies, not just sexual harassment, which in itself is bad enough). This means over 250,000 service members are estimated to have been victims since we uncovered this issue and brought it to the attention of the senior leadership of the Army and the Department of Defense.

The events that occurred affected my thoughts on leadership and helped me shape and communicate those thoughts to the several hundred officers, warrant officers, noncommissioned officers, soldiers, and civilians I have had the privilege of working with since leaving Aberdeen and the Army. The lessons I learned are applicable to today's leaders both in and out of the military. I include these lessons in my presentations on leadership to private sector business leaders.

The military services are rife with acronyms and abbreviations, and there are many within this book. I have provided a list of the most common abbreviations and acronyms for your reference and assistance. For the military acronym purists, please note I have taken some liberties with capitalization. I have also spelled out each abbreviation and acronym the first time it is used in each chapter for ease of reference.

I have also provided a list of principals to help identify the many people and organizations within and outside the Army who played roles in this disturbing chapter in Army history. This is, however, not a complete index of all the names used in the book.

To protect the privacy of the eight principal trainees involved in the Aberdeen cases, we did not use victims' names. We identified the perpetrators, under my command, by name only if their names appeared in official US Army press releases.

1

★ ★

MY NEW ASSIGNMENT BEGINS

FRIDAY, AUGUST 11, 1995, WAS A BEAUTIFUL, YET TYPICALLY muggy, summer morning at Aberdeen Proving Ground in Maryland. I met with the current commander of the US Army Ordnance Center and School (USAOC&S), Major General (MG) James W. Monroe, in his office in Simpson Hall to review the sequence of events for our change of command, which was scheduled to take place on Fanshaw Field at 1000 hours. I was a brigadier general, a one-star not yet selected for promotion, and MG Monroe was a two-star. He and his wife, Charlyne, had been most gracious to my wife, Ellie, and me throughout the transition process. The ceremony was superb—several hundred soldiers in formation and all the flags, cannons, band, and Army equipment on display.

In 1995–96, the Ordnance (Ord) Corps was the largest branch in the Army with 129,000 officers, warrant officers, noncommissioned officers (NCOs), and soldiers in the Active Army, National Guard, and US Army Reserve spread all over the world. This represented 13 percent of the total Army. There were 3 officer areas of professional concentration, 12 warrant officer military occupational specialties (MOSs), and 52 enlisted MOSs—and we trained them all.

Just about every unit in the Army had Ord soldiers who provided munitions, maintenance, or explosive ordnance disposal (EOD) support. The ethnicity profile of the Corps mirrored that of the Active Army, but there was a lower percentage of female soldiers in Ord compared with those overall in the Active Army. The lower percentage of females was determined to be because the MOSs of Ord—such as mechanics and welders—were those most often associated with male professions in the civilian workforce.

As the chief of Ordnance, I was responsible for all matters related to the Corps in regard to doctrine, training, leader development, organization, materiel, and soldier issues for all 129,000 Ord soldiers and leaders.

With regard to training execution, we trained an average of 25,000 students per year at 11 installations: Redstone Arsenal, Alabama (RSA); Eglin Air Force Base, Florida; Fort Gordon, Georgia; Fort Knox, Kentucky; Aberdeen Proving Ground, Maryland (APG); Edgewood Arsenal, Maryland (EA); Indian Head, Maryland; Keesler Air Force Base, Mississippi; Fort Sill, Oklahoma; Fort Jackson, South Carolina; and Fort Bliss, Texas. (Note: For readers interested in but not familiar with the Ordnance Corps and initial entry training, refer to Terms, Definitions, and References at the end of this book.)

I commanded both the US Army Ordnance Center and School (USAOC&S) at APG and the US Army Ordnance Missile and Munitions Center and School (USAOMMC&S) at RSA.

The USAOC&S consisted of the 61st Ord Brigade (Bde) and was headquartered at APG, along with its subordinate, 16th Ord Battalion (Bn). The second battalion, 143rd Ord Bn, was located 10 miles away at EA. These two battalions provided command and control for the students in training. Students included marines, sailors, and airmen, as well as Army soldiers.

The USAOMMC&S consisted of the 59th Ord Bde and its 832nd Ord Bn headquartered at RSA and the 73rd Ord Bn stationed at Fort Gordon, Georgia. As at APG, students from all four services were trained at RSA, as were civilian law enforcement personnel in EOD.

While I commanded both of the schools, I was responsible for only the content of the Ordnance course instruction at Forts Jackson, Sill, and Knox.

While the traditional headquarters for the chief of Ordnance was at APG, I was also expected to spend 2 to 3 days a week at my other office at Fort Lee, Virginia, approximately 210 miles south of APG. I was also the Deputy Commanding General for Ordnance at the US Army Combined Arms Support Command (CASCOM) headquartered at Fort Lee.

While still supporting operations in the Balkans and continuing to meet other strategic commitments around the globe, the Department of Defense in general and the US Army in particular were in the midst of efforts to reduce the Defense budget through efficiencies and reorganizations in 1995. This was all in an attempt to realize more of the anticipated "peace dividend" following Desert Shield/Storm. These budget reductions significantly affected the situation I experienced at APG.

In essence, we were renters, not owners, of the facilities we occupied. Therefore, I had neither command of all personnel on APG and RSA (for example, the military police [MP], staff judge advocates [SJAs—lawyers], and medical personnel) nor control of the installation budget.

Prior to my assumption of command, I had several discussions with MG Monroe regarding his assessment of the status of the command. I also spoke with three former chiefs of Ordnance to obtain their advice. I also discussed my upcoming duties with my soon-to-be direct boss, MG Thomas W. Robison, commanding general (CG) of the US Army Combined Arms Support Command (CASCOM) at Fort Lee, Virginia, and his boss and my senior rater, General William W. Hartzog, CG of the US Army Training and Doctrine Command (TRADOC) at Fort Monroe, Virginia.

I also consulted with chiefs of other branches in the Army who had done or were doing the same type of functions I would be doing, and I contacted several old friends who had served in various leadership positions at APG.

Following the change of command reception, I met with the senior leadership of both the USAOC&S and the USAOMMC&S in the command conference room in our headquarters in Simpson Hall to go over my command philosophy.

My guidance to the leadership covered 15 points, with 3 being of particular importance to the situation with which we were about to be confronted. I wanted everyone to (1) set and enforce the highest technical, tactical, moral, and ethical standards for staff, faculty, and students; (2) keep me informed; and (3) support 4 major programs: (a) prevention of sexual harassment; (b) prevention of driving while intoxicated offenses; (c) prevention of drug abuse; and (d) support of a positive equal opportunity/equal employment opportunity environment.

I then spent several weeks touring the training facilities under my command, meeting with the leaders, meeting with the leadership of the installation at which my organizations resided as tenants, and talking with the young soldiers in training.

During November 13–16, 1995, Colonel (COL) Charles E. Beckwith, the TRADOC inspector general, his sergeant major, and a team of 5 other inspectors visited USAOC&S at APG and EA to assess the command climate, among other things. This visit was within a few days of the 90-day anniversary of the change of command. COL Beckwith's team conducted 17 interviews and 13 sensing sessions involving more than 200 individuals.

They found the command climate at USAOC&S was positive in all areas. Soldiers and civilians were supported by a professional and caring chain of command. Although leaders at all levels expressed concern over the numerous problems associated with diminished resources, they continued to accomplish their missions through creativity, hard work, and dedication to the workforce. COL Beckwith also concluded there was a functioning organization inspection program—all soldiers were familiar with the command philosophies and policies. Risk management and risk assessments were being done in every instance.

All of these actions constituted what is referred to as the incoming commander doing a command climate assessment upon change of command. In 1995, there was no requirement by the Department of the Army for a new commander to do such an assessment. As a result, there was no format or guide for a new commander to follow.

At the end of my first 90 days in command, I felt I had a pretty good feel for the command and the areas that needed work.

First, I needed to move Russ Childress, the senior civilian at USAOC&S, back into the command group. He had been relegated by MG Monroe to a position in the training department. It was important to have our more than 2,000 civilian employees represented when decisions were being made, and their representative needed to have immediate and free access to me.

Second, it was important for me to visit the parts of our organization in the 9 locations outside of the APG/EA area and the units in which the Ordnance soldiers we trained were serving the field Army around the world. As a result of budget cuts, we no longer had an organization that would visit units to determine if we were training our students on the skills they needed. I would have to make these visits, in addition to my extensive travel schedule dictated by the need to be at Fort Lee often and to attend the large number of conferences, boards, meetings, etc. required of senior leaders.

Third, I needed to understand and assess the CASCOM reorganization, which had been implemented the year before to save money by reducing people, and I needed to try hard to make it work. The cuts were massive and difficult, but the workload continued and expanded. As a result of the organizational challenges forced upon our schools, many folks had to perform additional duties and come up with work-arounds to get their jobs done. However, we seemed to be doing all right; most commanders in the field reported they were generally satisfied with our graduates.

The peace dividend may have looked good in Washington, DC, but for us at APG, it was not a pretty picture.

In addition to the daily task of training Ord personnel of all ranks, we also focused on several key initiatives that would increase our soldiers' ability to better support the field Army. These efforts included finalizing the Ordnance Corps Vision document to be sure we were aligned with the Army's vision and developing tele-maintenance to provide mechanics in the field with access to experts in the schoolhouse—as the medical community was doing with tele-medicine. We also were expanding the use of prognostics and diagnostics to increase the probability of accurate and early problem identification to reduce the time and cost of repairs. We also began consolidating MOSs to reduce training costs.

To help our soldiers in the field, we were consolidating, rewriting, and publishing a streamlined set of Ord field manuals. We were also transitioning from paper to electronic technical manuals used by soldiers to guide them in their repair work.

On the equipment side, we were ensuring the proper materiel was being developed for the Ord soldier: improved mobile maintenance vehicles, a new heavy recovery vehicle, and test measurement and diagnostic equipment, to name a few.

Because both internal and external assessments indicated we were on the right track, I informed the leadership at USAOC&S and USAOMMC&S of my decision that we would operate as the CASCOM reorganization envisioned for a year. We would assess how it was going as we went along and then make any major changes required during the second year of my command tour.

I felt we would have to document problems in order to regain any of the 1,000-plus staff and faculty and $34.4 million in budget per year Ordnance had lost in this well-intentioned, but flawed, attempt to realize savings. There was no way I could go to the Army senior leadership after being in command fewer than 4 months and say, "This CASCOM reorganization is not good, and we need more resources." We had to first see if we could make it work.

On February 20–21, 1996, we conducted a leadership review of our progress at the 6-month mark of my command tour at RSA. The team had open discussions, and we identified several areas to work on to improve operations. As expected, most of the problems centered on resources and the lack thereof.

Major (MAJ) Susan Gibson, the deputy staff judge advocate at APG, and MAJ Phil Hartsfield, the APG provost marshal went over the USAOC&S review and analysis (R&A) data for the first quarter of fiscal year 1996 (which ended on December 31, 1996). They highlighted what appeared to be an increase in alcohol-related incidents of misconduct by trainees off-post in the local area.

During analysis of the second quarter R&A data in April 1996, the Aberdeen Criminal Investigation Command (CID) Resident Agency pointed out that the data indicated an increase of trainee-on-trainee sexual misconduct, which appeared to be alcohol-related in most cases. Young soldiers were pooling their money and renting motel rooms in one of the surrounding communities. Invariably, the drinking would result in males and females engaging in sexual activity that, in some cases, was not voluntary by the female.

We had no comparison data available from other units or installations, even though MAJ Gibson tried several times to get baseline data from multiple sources so we could determine whether we had a problem or whether we were experiencing a normal level of such activity. The only information we had was trend data about ourselves.

Based on what we were seeing with regard to underage drinking and trainee-on-trainee sexual misconduct, we obtained assistance from the CID, SJA, and the MPs to work on the underage drinking problem. We also directed even more NCO presence in the barracks, including moving drill sergeants into the troop barracks to provide more direct supervision. The commanders worked to reduce the alcohol-related incidents for several weeks. The 61st Ord Bde commander also initiated an intensified rape prevention program. An almost immediate decrease in incidents of alcohol-related sexual misconduct among trainees was noted.

2

★ ★

OUR SOLDIERS REPORT
SEXUAL MISCONDUCT

SPRING AND SUMMER ARE THE BUSIEST TIMES FOR CHANGES of command and personnel moves in the Army.

On April 22, 1996, Colonel (COL) Thomas A. Hooper relinquished command of the 59th Ordnance Brigade (Ord Bde) at Redstone Arsenal to COL Jerry Luttrell. Tom had done a wonderful job. He and his wife, Diana, were a superb command team. As a result of COL Luttrell's departure as my deputy commandant at Aberdeen Proving Ground (APG), I made Russ Childress the acting deputy commandant until Lieutenant Colonel (LTC) Johnnie L. Allen arrived to replace COL Luttrell.

Major General (MG) Ken Guest became commanding general (CG) of CASCOM and my direct boss.

Immediately after that, LTC Donald J. Hogge became the new commander of the 16th Ordnance Battalion (Ord Bn) at APG, and COL Dennis M. Webb assumed command of the 61st Ord Bde at APG. My new team was in place.

Within 10 days of taking command, COL Webb became aware of

suspected sexual misconduct involving drill sergeants (DSs) and noncommissioned officer (NCO) instructors with female trainee students.

I was on temporary duty on a selection board in Washington, DC, between July 8 and August 7. COL Webb kept me informed by phone and email. He informed me that he had opened an investigation of the alleged misconduct under the provisions of Army regulation (AR) 15-6.

AR 15-6 investigations gather and consider all relevant evidence about an incident, possible misconduct, or a failure to adhere to regulations or policies by personnel under the appointing authority (AA). The investigating officer (IO) must be thorough and impartial and consider all sides of the issue. The IO must submit a written report to the AA containing findings and recommendations that must be based on the evidence discovered and must comply with the instructions of the AA. The IO's report should answer the following questions: what happened, when did it happen, where did it happen, who was involved, and why did it happen.

I told COL Webb I was concerned, charged him to dig into the situation, and requested he keep me informed. COL Webb later provided the following summary:

> On Tuesday morning, 9 Jul 96, the commander of the 143rd Ordnance Battalion at Edgewood Arsenal informed COL Webb that he suspected an inappropriate relationship between a drill sergeant, Staff Sergeant (SSG) Nathaniel C. Beach, and a female trainee, both assigned to the battalion's C Company (C/143rd) at Edgewood Arsenal (EA). The C/143rd first sergeant had found a letter written from the trainee to SSG Beach, and the trainee subsequently admitted she had had sexual contact with Beach. The battalion commander immediately suspended SSG Beach from drill sergeant duty, imposed a no-contact order on him, and reassigned

him to Aberdeen (APG) pending further action. COL Webb then appointed a former Advanced Individual Training company commander to go down to Edgewood and conduct . . . a 15-6 investigation. Through this investigation COL Webb was able to substantiate the allegations against SSG Beach. During the out-briefing by the investigating officer, he indicated that other cadre members' names had come up during the interviews with trainees as suspects for other inappropriate relationships. Two other cadre members were subsequently identified as involved in misconduct. The 143rd Ord Bn commander and staff worked with the Staff Judge Advocate to develop the legal case in these matters.

COL Webb sensed some underlying factors could have precipitated the misconduct. He requested the Defense Equal Opportunity Management Institute (DEOMI) at Patrick Air Force Base, Florida, perform a Military Equal Opportunity Climate Survey (MEOCS) of the entire 61st Ord Bde, including civilian employees. This was the only assessment tool, to our knowledge, to help us determine if there were systemic problems.

During the latter part of July, LTC Hogge informed COL Webb he suspected similar activities had occurred or were occurring in his new command, the 16th Ord Bn at APG.

The AR 15-6 investigation ran until late August, when the investigation was turned over to the installation provost marshal, because some of the suspects of possible misconduct were reportedly threatening trainees not to inform on them.

On August 23, I was frocked to the rank of MG. This meant that while I was authorized to wear the two-star insignia of rank and receive the honors and courtesies extended to a MG, I would continue to be paid as a brigadier general (BG) until I was officially promoted. General

(Retired) Jimmy D. Ross officiated at the ceremony. General Ross had promoted me to COL and BG, and I was honored that he and Mrs. Ross were able to share the day with Ellie and me.

I departed for Germany on September 5 to begin a weeklong trip in Europe to visit Ordnance soldiers, speak at the Annual Ordnance Ball in Germany, and visit my counterpart in the German Army as part of an exchange program.

Back at EA on September 5, a trainee reported that her DS, SSG Delmar G. Simpson, had attempted sexual relations with her. She reported this sexual misconduct allegation to her company commander, Captain (CPT) Derrick A. Robertson. He immediately counseled the trainee on the implications of false swearing and then took a written sworn statement from her. CPT Robertson started an inquiry into the allegation and notified his chain of command. He continued to talk with soldiers in the company and investigate further. During his investigation, CPT Robertson uncovered the names of more of SSG Simpson's alleged victims.

On September 6, CPT Robertson notified the Bn chain of command of the allegations against SSG Simpson, and he ordered SSG Simpson to remain in his quarters and not have any contact with soldiers until further notice. The acting commander of the 143rd Ord Bn called COL Webb to request an AR 15-6 IO from outside the Bn. COL Webb appointed CPT Sheila Bruen, the commander of A Company, 16th Ord Bn at APG. CPT Bruen had prior service as an enlisted soldier before becoming an outstanding officer.

The following Monday, CPT Bruen collected all the data and statements from CPT Robertson and contacted the installation staff judge advocate's (SJA's) office for legal advice. She began to review and analyze the data and formulate a plan for the next day's activities.

Throughout the next day, CPTs Bruen and Robertson worked closely together to determine the who, what, when, where, and how of the situation. They were notified that the US Army Criminal Investigation Command (CID) was also investigating SSG Simpson for the same allegations. At that point, the investigation had also revealed the possibility of other DSs being involved in sexual misconduct.

On September 11, the acting commander of the 143d Ord Bn at EA informed COL Webb of the involvement of CID. After consulting with legal, it was determined that the AR 15-6 investigation being conducted by CPT Bruen should be terminated so as not to interfere with the CID investigation. The military police (MP) were told to apprehend SSG Simpson and begin to process him for pretrial confinement. Considering the seriousness of the allegations and the fear of reprisal expressed by the trainees, an order from a military judge was obtained. SSG Simpson was apprehended, charged, and transported to the US Marine Corps detention facility at Quantico, Virginia.

CPT Bruen later provided me with this recollection of her time as the AR 15-6 investigating officer:

> Part of the 15-6 package that I received contained several sworn statements from female soldiers going through Advanced Individual Training (AIT) who alleged that they had been sexually abused by SSG Simpson. I then made arrangements to interview the females who had made these statements. I was shocked by their allegations. Having been a private during the early part of my career, I knew the relationship between trainees and drill sergeants is built on trust, respect, and a little fear. It saddened and sickened me to think that this trust had been horribly abused. If these stories were true, I felt these young women had been manipulated and abused by the very ones who were supposed to protect them. In my company, I always made a point of telling newly arrived soldiers that they could talk to me whenever they felt that they had a need. After meeting with these young women, I immediately determined that their statements warranted further investigation. Shortly thereafter, I was told that the 15-6 investigation had ended due to the involvement of CID.

While investigating the Simpson allegations, a trainee in C Company, 16th Ord Bn at APG informed the CID that she had been raped by her DS, SSG Wayne A. Gamble. We now had two DSs in two organizations at two locations (EA and APG) suspected of raping trainees. My trusted aide, CPT Jerry D. Stephens, called me in Germany on September 11 and informed me of the worsening situation.

I immediately thought of how the leaders I had worked for in the past had handled really bad situations. The guiding words of one of my old bosses, Admiral Paul David Miller, were "Do right." The commander must not worry about what may happen to him or her; the commander should only worry about taking care of the people under his or her command and doing the right thing.

I returned from Germany on September 13 and landed at Dulles International Airport in Washington, DC, at 1630. CPT Stephens was there with my driver to pick me up. COL Webb lived two doors down the street from us in Quarters #8 and was waiting on the steps of Quarters #10 when we pulled into our driveway. Being over 6 feet tall, riding in coach always physically taxed me, especially on international flights, and I was really beat.

After dropping my bags and greeting my wife, Ellie, and Remington (our 10-year-old Shih Tzu), COL Webb and I headed to the sun porch so he could update me on the situation. I directed we set up a crisis action team that included SJA, CID, and the MPs. I said we would do three things: (1) act on the current situation; (2) identify systemic causes; and (3) institute corrective actions.

After COL Webb left, I went into the kitchen, where Ellie was sitting with Remington so he wouldn't bother us. When I walked into the kitchen, she asked, "What's wrong? You look white as a sheet!" I told her what was going on and how bad the situation was. It doesn't get any worse than leaders abusing their subordinates. She asked me what I was going to do about it, and I told her the guidance I had given COL Webb. She wisely said, "That sounds all well and good, but the first thing you have to do is take care of those women," referring to those who had been victimized.

Ellie always gave sound advice. She was well read, she truly understood and liked people, and she had had an experience few others have had. I met her when I was a student at the US Army Command and General Staff College in Fort Leavenworth, Kansas, in 1977. She was the administrative assistant to the manager of in-flight services for Trans World Airlines (TWA) at the Kansas City International Airport.

A friend of mine was married to a TWA flight attendant, and they thought Ellie and I would be a good match. They set up a blind date, and I was waiting with LTC Merrill Steele in their house in the city of Leavenworth when his wife, Nancy, brought Ellie home with her.

That night at an art auction, I learned about Ellie's background. She was born and raised in Lincoln, Illinois, and after graduating high school, she went off to college for two years, but decided to pursue a career as a flight attendant. Ellie flew for TWA for 10 years and really enjoyed it, meeting all sorts of interesting people on her flights.

On the night of November 20, 1967, she was working flight 128 from Los Angeles to Cincinnati and then on to Boston with another stop in Pittsburgh. She was sitting in the very back of a Convair 880 (a 4-engine jet that looks somewhat like a Boeing 707) on a jump seat opposite the galley. While attempting a night landing at the Cincinnati airport in a light snow fall and with the approach lights out at the airport, the pilot apparently mistook the lights on the Ohio River as ground level while landing to the southeast on Runway 15. They were almost 500 feet too low, and the plane hit a large tree on the south bank of the river. The impact sheared off the cockpit, and the plane crashed and burned in an apple orchard. Ellie was one of the 13 initial survivors (only 10 would ultimately survive) out of a total of 82 passengers and crew on board.

As a result of my discussions with Ellie about the allegations, I came to the conclusion that the three actions I had relayed to COL Webb remained valid, but our objectives needed to be threefold: (1) identify potential victims and ensure we provide all necessary support to them; (2) identify alleged perpetrators and allow the judicial system to work its due process; and (3) identify systemic causes for the problem and initiate corrective actions to preclude recurrence. These three "vectors" guided me and the actions of our team for the remainder of my involvement with the Army's Aberdeen sex scandal.

After my discussion with Ellie, I called COL Bob Wilson, the executive officer for General (GEN) William W. Hartzog, Commanding General (CG) of the US Army Training and Doctrine Command (TRADOC) at Fort Monroe, Virginia, to let him know what was going on. I asked him to pass the information on to GEN Hartzog.

The next morning, I called my immediate boss, MG Guest, who was on travel, and told him about the situation that was unfolding and about my call to GEN Hartzog's office the night before. We talked about getting some outside eyes and ears to help us get to the bottom of how and why this happened as we worked through the legal actions. He agreed the TRADOC inspector general (COL Beckwith) would be a good person because he and his team had been at APG in November 1995 and TRADOC headquarters was two levels of command above me.

3

★ ★

THE "GAME" IS DISCOVERED

On September 15, 1996, I met with Colonel (COL) Dennis M. Webb and Major (MAJ) Susan S. Gibson, the installation deputy staff judge advocate (SJA), and told them we were getting the inspector general (IG) from Training and Doctrine Command (TRADOC) to help us work the problem. I confirmed our three-pronged attack and objectives.

COL Roslyn Glantz, the Garrison commander, had recommended to COL Webb that we put out a press release on the Simpson case. Timing of press releases became a major daily decision topic. We decided to wait as long as possible to keep from interfering with the numerous ongoing investigations and to help avoid even the appearance of command influence.

I signed and faxed a memorandum to General (GEN) William W. Hartzog confirming my calls to his staff and officially requesting "that you appoint an independent investigating officer to conduct a complete investigation of the command climate and unit circumstances leading up to and surrounding the recent allegations of rape, sexual assault, and improper cadre/trainee relationships at the U. S. Army Ordnance Center and School. Criminal Investigation Command (CID) is thoroughly investigating and we are taking appropriate legal action. However, if there are systemic problems in the command that fostered these illegal acts or inappropriate relationships, we need to get to the root of those problems and take corrective action."

On Monday, September 16, I called GEN Johnnie E. Wilson, Commanding General (CG) of the US Army Materiel Command to give him a heads-up that we were working on a drill sergeant (DS) sexual misconduct problem. GEN Wilson was the senior Ordnance (Ord) officer on active duty, and the installation known as Aberdeen Proving Ground (APG) was under his command.

Later in the day, I was informed by GEN Hartzog's chief of staff, Major General (MG) James J. Cravens, Jr., that the TRADOC IG would not be coming, but instead COL Raymond L. Rodon, commander of the 23rd Quartermaster Brigade (Bde) at Fort Lee, Virginia, would be the 15-6 officer.

That evening, Ellie and I attended a farewell dinner at the Top of the Bay Club at APG, a neat stone building with a fantastic view of the Chesapeake Bay that had at one time been the Officers' Club. Partway through the meal, I received a note saying that Chief Warrant Officer 3 (CW3) Don Hayden needed to see me urgently outside. CW3 Hayden was the agent in charge of the APG CID Resident Agency.

CW3 Hayden said, "Sir, you are not going to believe this."

I said, "Don, I'll believe about anything by now. What have you got?"

He went on to relate that the day before, a trainee reported that Captain (CPT) Derrick A. Robertson had invited her to his quarters where he subsequently raped her. She was the same trainee who had first accused Staff Sergeant (SSG) Delmar G. Simpson. CW3 Hayden read her statement to me, and it sounded like an instant replay of the one she had submitted accusing SSG Simpson.

Now we had the captain who initiated the action on the alleged misconduct by SSG Simpson and had escorted SSG Simpson to the Marine Corps Brig doing the same thing. It was unbelievable that a captain in the Army would do such a terrible thing to one of his subordinates.

CW3 Hayden and I convened a meeting away from the party attendees, with COL Webb and COL Edward W. "Buzz" France, the installation SJA, to review the Robertson situation. CW3 Hayden informed us that the CID was also preparing a case on another DS, SSG Vernell Robinson, for allegedly engaging in consensual sex with a trainee in violation of one of our USAOC&S regulations. SSG Robinson had been in the same company of the 16th Ord Bn at APG as another DS suspect, SSG Wayne A. Gamble, before SSG Gamble was transferred to Fort Bragg, North Carolina, in August.

Later that evening, I called GEN Hartzog and MG Ken Guest and informed them of the growing number of sexual misconduct cases.

I was informed that CPT Robertson admitted to having what he felt was consensual sex with a trainee. I immediately concurred that CPT Robertson should be relieved of command. He was replaced that day by CPT Alicia Jackson. This made great sense to me because Jackson was a superb officer and we could certainly use a woman's perspective on what was happening within that company.

Before departing for Fort Monroe on September 17, I signed the appointment memorandum to COL Rodon informing him that a CID investigator had been assigned to help him and that MAJ Gibson would be the first person he talked with to ensure his investigation coordinated with ongoing legal actions. I directed him to address eight specific areas I wanted covered by his investigation.

I called the CG of the 1st Corps Support Command at Fort Bragg, North Carolina, and alerted him to the ongoing investigation of SSG Gamble. I asked him to be sure SSG Gamble was not engaging in the same misconduct in his unit at Fort Bragg.

That afternoon, I stopped by GEN Hartzog's office at Fort Monroe and gave him a face-to-face update on the situation. He reiterated that I was in charge and that I should make sure I proactively corrected any problems uncovered by the inquiry.

On the way back to APG, I made up my mind to expand the attendance at the daily crisis action team (CAT) meetings, increase the frequency to twice a day, and chair every meeting either in person or by phone if I was not at APG. I felt I needed an up-to-date status on what CID and military police (MP) investigations were uncovering and on the corrective actions that were evolving and being implemented. I had learned over the years that in a tough situation, the leader should go to the point of the action and become personally involved.

After discussing possible public affairs implications and the status of the investigation at the September 18 update, I directed that we all needed to avoid turning this situation into a witch hunt. The objective was to find the female trainees who had been abused and from their statements identify any potential perpetrators. We would let the judicial system work, and we would avoid even the appearance of command influence. We would also be using what COL Rodon and his team uncovered to initiate corrective actions to preclude recurrence.

I then announced our CAT would meet each weekday at 0730 and 1700, each Saturday at 0730, and each Sunday at 1700 and that I would be chairing each meeting either in person or on the phone.

Later in the day on September 19, I updated MG John M. Longhouser, the new APG installation commander, on the unfolding events. I thanked him for the great support I was getting from the installation staff. We also agreed that because he was the general court-martial convening authority (GCMCA) for APG and Edgewood Arsenal (EA), I would not pass on any information to him with regard to cases. I believe I was the only TRADOC general officer (GO) school commandant who was not a GCMA.

The one advantage this provided is that I could get details of ongoing legal actions and use this information to effect corrective actions without technically being concerned about command influence. In reality, I was very concerned about command influence by the mere fact that I was a GO and what I said verbally or with body language could influence people. I did not want to jeopardize the investigative and subsequent legal processes.

COL Rodon came to see me with results from his first day of investigation. The key points for the day were how the lack of chaplains not only affected the flow of information, but also precluded an effective counseling program for DSs and their spouses; the equal opportunity program was not effective; and the command team in C Company, 16th Ord battalion (Bn) was weak.

The next morning, COL Webb informed me an alleged trainee-on-trainee rape had occurred the previous evening. A male and a female trainee went to the on-post Burger King after attending sensing sessions with COL Rodon's investigating team. As the trainees went back to their billets, they starting kissing, and the male continued his actions after the female said to stop. She reported to her DS that she had been raped, and he reported the incident to the MP.

The male trainee later admitted to the rape. Right in the middle of all the attention we were putting on preventing sexual misconduct, and almost immediately after a sensing session about sexual misconduct, we had this happen. It was also troubling that the female soldier was apparently not with her female battle buddy. Female soldiers were to go everywhere in pairs to help preclude exactly what had happened.

I then met with our senior leadership and stressed again the importance of leadership in the billets, enforcement of the buddy system, and following policies and procedures.

The one good thing about the previous evening's incident was that the female trainee felt confident something would be done, so she reported it to her DS, and he took the correct action. This case was another reminder to me that our soldiers had confidence in the senior leadership and that they were the ones who brought forward the misconduct, not some outside investigator.

It would become a matter of routine that I would call GEN Hartzog or someone on his staff, as well as call my immediate commanding officer, MG Guest, at least once a day to keep them informed as we worked through the crisis. I imagined the dread they felt every time their staff announced I was on the phone and wanted to talk with them.

MG Joe Bolt, CG of the US Army Training Center at Fort Jackson, South Carolina, called to offer his help. I suspected someone at TRADOC headquarters had called him and advised him of the situation at APG. Fort Jackson was one of the initial entry training installations where both basic training and advanced individual training (AIT) was conducted, and it was also an installation where DSs were trained. I later sent him an email thanking him for his call and advising him that the "15-6 is still ongoing, but believe I'll need your assistance in correcting any problems uncovered on the 'technical side' of the drill sergeant business . . . just a heads-up for now."

COL Rodon reported that trainees were becoming reluctant to talk. He also highlighted the problem with holdovers and reclassified students being more likely to be involved in misconduct as they had more free time on their hands.

Holdovers were students who had graduated but were waiting completion of security clearance processing and/or resolution of medical issues that precluded them from moving on to their first duty assignments.

Reclassified soldiers were individuals who had been in another branch of the Army and were now in Ord. In most cases, they had been through basic and AIT somewhere else and were now at US Army Ordnance Center and School (USAOC&S) to get an Ord military occupational specialty.

At the evening update, I announced I wanted to start collecting basic data on each subject and victim to see if any common threads would help focus our corrective actions. My industrial engineering background was coming through. This effort started slowly, but over the next few days, we developed a series of spreadsheets that proved invaluable to me for keeping all the cases straight and analyzing trends.

At the morning update on Saturday, September 21, CW3 Hayden noted that two potential victims may have gone absent without leave (AWOL) back in July. This raised the question of whether we should go back and contact all soldiers who had gone AWOL to see if sexual misconduct by a DS or an instructor was the reason for their action. We decided to investigate female soldier AWOLs.

Later that day, I talked with COL Cecily David, the commander of Kirk Army Health Clinic at APG. The clinic had been a hospital until budget cuts and efficiency initiatives resulted in its downgrade to a health clinic. COL David was a caring physician and leader. We talked about the support that needed to be provided to the victims. I invited her to join our CAT meetings starting the next Monday morning, and she accepted. This turned out to be a very smart decision and one I should have thought of a week earlier.

It also struck me that we needed to have public affairs experts in our meetings because we would eventually need to put out a formal press release. This also turned out to be a wise decision, and the team expanded to include both Gary Halloway, the installation public affairs officer (PAO), and Ed Starnes, the Ord Corps PAO. Ed was a real pro—a big man with an even bigger brain and heart.

Also on September 21, I asked CW3 Hayden and his CID team to take over all sexual misconduct investigations involving APG and EA personnel assigned to USAOC&S. I asked him to include not only the felony cases CID normally investigated but also the non-felony cases doctrinally done by the installation's MP. I wanted consistency in investigations and one belly-button to push, CW3 Hayden's, to find out what was going on.

We continued to work actions through the weekend, and at the 0730 CAT meeting on Monday, September 23, we had the team in place that would meet per the original schedule until October 14, when we reduced the frequency to once a day.

During the meeting, we noted it was apparent that some of the female trainee victims were having significant emotional problems and that mental health issues were becoming a major concern.

After the meeting, I met with COL Webb to go over the spreadsheets MAJ Gibson had prepared on the four major cases as of September

23. The day before, he and I had discussed examining past investigations and a range of other issues, including the need to increase leadership presence in the barracks.

At the 1700 nightly CAT meeting, MAJ Phil Hartsfield, the installation provost marshal, said he had learned that investigations of criminal misconduct were also ongoing at three other TRADOC training installations. This was the first time anyone had mentioned we were not the only ones with this type of problem. I had mixed emotions about this, but was personally relieved that at least I was not the only commander working such cases.

The team briefed me via phone on the morning of September 25 regarding cases being worked at other installations: Fort Monroe, Virginia (unspecified number); Fort Rucker, Alabama (1 case); Fort Huachuca, Arizona (10 rapes and 4 assaults); and Fort Leonard Wood, Missouri (12 rapes and 22 assaults). I later learned that Fort Leonard Wood was one of three installations mentioned the day before and that the other two were Fort Jackson, South Carolina, and Fort Lee, Virginia. That made six TRADOC installations with problems in addition to us.

The Public Affairs Office team reported that Paul Boyce, the CID Public Affairs Officer (PAO) at Fort Belvoir, Virginia, was being very helpful, and we needed to keep him updated on what we were doing regarding any announcements. Paul Boyce ended up being a trusted adviser to me personally and an invaluable resource for our team.

It was becoming more apparent every day that a press release would need to be published. In the meantime, we would maintain the position to "respond to query" only in order to preclude any possible interference with the ongoing investigations and legal actions.

I took away three things from the daily update with COL Rodon. First, in addition to all the other things we were looking at, we needed to analyze the cases of female soldiers who had been chaptered out of the Army. Some trainees may not have performed well or may have appeared to be unsuitable, not because they were bad soldiers, but because they were victims of sexual harassment or sexual misconduct by cadre at APG or EA. It appeared our rate of discharge was higher than one would expect; however, with no comparative data, it was just an observation when compared with COL Rodon's Bde data. Not having a TRADOC- or Army-wide database to reference was a problem for us.

Second, we needed to consider having a separate training company for holdovers, reclassified soldiers, and soldiers with prior service, as COL Rodon had in his Bde.

Third, we needed to reexamine the policy where AIT students were not permitted to smoke. This led to people sneaking into isolated places to smoke, making them targets for fellow trainees or susceptible to cadre offering to allow them to smoke in return for favors.

The nightly update was encouraging, as no new subjects had been identified. MAJ Gibson reported she had been informed by Lieutenant Colonel (LTC) Scott C. Black in the Office of the Chief of Legislative Liaison, Office of the Secretary of the Army in the Pentagon that he had been in contact with J. P. Sholtes, legislative assistant for Congressman Bob Ehrlich, Jr., 2nd District of Maryland, within whose district APG fell.

After the nightly update, I met with my deputy, LTC Johnnie L. Allen; my chief-of-staff, LTC Gabe Riesco; and our senior civilian, Russ Childress, to discuss changes we needed to make in the organization and in policies, based on the emerging results of COL Rodon's investigation. We determined we would implement all improvements under our control as soon as possible. I directed we establish a formal get-well action plan so we could document the problem found, the action taken, and the status. I wanted to get on with actions that might help preclude recurrence of the problems we were uncovering.

I then called MG Morrie Boyd, the chief of legislative liaison who reported to the secretary of the Army, about his office's contact with Congressman Ehrlich's office. I figured, if we were going to be responsible for what was reported to Congress, then we should have the authority to deal directly with the members and the staffs, as appropriate. Both Congressman Ehrlich and Senator Barbara Mikulski told me to call them personally at any time. Senator Paul Sarbanes's staff encouraged me to do the same when I had met with them in 1995.

At both updates on September 26, it was noted that the mental health of at least two of the alleged victims was very poor and that they needed psychiatric help. COL David was working to make that happen. One trainee's testimony would be needed in court, but she now did not want to testify.

MAJ Hartsfield reported at the evening update on his efforts to compile data on the more than 15 female trainees who had gone AWOL since January 1, 1995. We decided to reach out to each of these soldiers to see if their action was precipitated by sexual misconduct by cadre members. The number of former female trainees we needed to check on kept growing as we worked our primary objective: identify potential victims and ensure we provide all necessary support to them.

I ran with the 16th Ord Bn during their physical training session the morning of September 27 and was not impressed with what I saw or heard when I talked with folks. It was obvious some junior leaders did not understand the seriousness of the situation and did not have the sense of urgency needed to affect change. I informed COL Webb that I wanted to see him, his CSM, and both Bn commanders and their command sergeants major (CSMs) at 1130 on Monday, September 30.

After the Friday morning update, MAJ Gibson met with J. P. Sholtes from Congressman Ehrlich's office. She later reported that he had agreed with what we were doing and how we were handling the situation.

After the Saturday morning update, I called MAJ Pamela J. Royalty, chief of Mental Health Services at Kirk Army Health Clinic, to check on the status of the two victims who required psychiatric care. One had been abused as a child, and this was bringing back bad memories for her.

In our conversation, MAJ Royalty brought up Playing the Game. She said this was a contest among DSs to see who could have sex with the most female trainees. This was the first I had heard of something like this going on, and it appeared that we were, in effect, dealing with a crime ring. This would play out over the next few weeks and months and would become one of the more sordid aspects of an already bad situation. I invited MAJ Royalty to join the CAT meetings from then on.

After my discussion with MAJ Royalty, I sent a note to COL Webb asking him to consider moving the office of the chaplain, who was assigned to the school as an instructor to teach ethics and morals, from the Bde headquarters area to the 16th Ord Bn. This would allow him to be immediately available to work with soldiers when not on the platform teaching.

On Monday morning, I called both MG Cravens and MG Guest to provide them an update and to advise them about Playing the Game. When I called MG Bolt at Fort Jackson, he said he knew about Playing the Game. I later learned "Playing the Game" was also the title of an early 1990s sexual assault prevention video.

I thought this was a big deal and would be of importance to other commanders who were also working on sexual misconduct problems. I could not imagine this was isolated only to DSs at APG and EA, as there were indications that DSs at basic training sites may have been passing to their DS buddies at AIT installations the names of female trainees who volunteered to give sex for favors.

The unraveling of the true circumstances with regard to players in the Game would eventually lead to my understanding of what really happened in the APG sex scandal and most likely the Army-wide sex scandal.

At the meeting with the Bn and Bde senior leaders, I expressed my total frustration with their lack of urgency. I always tried to maintain my calm, but at this meeting I used a lot of four-letter words. I stressed that taking care of soldiers and solving this problem was a big deal. This is the first time I used the expression: "This is the worst thing I have ever seen in my military career." A few of them in attendance had not seen me mad. Now they had.

LTC Dale Archer, commander of the Washington District of CID, visited CW3 Hayden's team, and I invited him to sit in on our nightly update. This became a standing operating procedure with visitors.

Anytime we had visitors from outside of APG/EA visit us about the situation, we invited them to sit in our updates so they could see firsthand how we exchanged information and made decisions.

As we closed out our second week of almost total immersion into sexual misconduct, I was getting more and more troubled about what we were uncovering. I was comfortable, however, that we had the team in place to work the issues and attain our three objectives.

It was very time consuming to have so many people tied up in meetings twice a day. But it facilitated the sharing of a range of views, allowed the cross-leveling of information, ensured continuity and unity of purpose, and, most important, provided me the understanding of what was going on so I could make sound and timely decisions.

4

★ ★

THE PROBLEM GROWS

O N OCTOBER 11, 1996, I LEFT FOR FORT CAMPBELL, Kentucky, to visit Ordnance (Ord) soldiers in the 101st Airborne Division and other, non-divisional units. I would also present a briefing on the state of the Ord Corps at a professional development class for a large group of Ord personnel of all ranks.

Colonel (COL) Dennis M. Webb reported by phone that Chaplain Moore would be reporting to the 16th Ord Battalion (Bn) the next day. COL Johnnie L. Allen reported that he had a very good session with Paul Boyce, Criminal Investigation Command (CID) public affairs officer, who had spent the day at Aberdeen Proving Ground (APG). Paul concurred with holding off on putting out a press release and agreed we should be responding to inquiries only.

The next morning, I learned that Brigadier General (BG) Daniel A. Doherty, commanding general of the CID at Fort Belvoir, Virginia, had briefed General (GEN) Dennis J. Reimer, chief of staff of the Army (CSA), and Sara E. Lister, assistant secretary of the Army for Manpower and Reserve Affairs, on the situation at APG and Edgewood Arsenal (EA).

Major (MAJ) Susan S. Gibson expressed her growing concern about command influence with me being so deeply involved. I stressed that all legal actions must be thoroughly worked with the staff judge advocate (SJA) and that in no way did I want to influence their action. From the very beginning of our daily meetings, all sensitive discussions regarding the investigations being conducted by CID or legal actions being worked by SJA were held in a stay-behind with only MAJ Gibson and Chief Warrant Officer 3 (CW3) Don Hayden and me in attendance. This meant that none of my subordinates would be involved in discussions of cases with me.

MAJ Gibson reported at the nightly update that the legal community was recommending that victims and subjects in investigations be attached to the APG Garrison for administration. This would thereby take our Bn commanders, COL Webb, and me out of the chain of command that would determine recommendations on legal actions. This would help to further reduce even the appearance of command influence by either my subordinates or me.

The announcement of our initial changes in the organization structure at APG/EA occurred on October 3. In an email to Major General (MG) Ken Guest, my immediate superior, I reported I had moved the training departments out from under the Bns at APG and EA and put them under the director of instruction, thereby freeing up the Bns to focus solely on soldierization (transitioning individuals from civilian to military life). I also let my boss know we moved the office of the chaplain from the Command and Staff Training Department to the 16th Ord Bn so he could be more readily available to help soldiers.

During the morning update on October 4, CW3 Hayden announced CID would be going back three classes to interview female trainees. This period would cover the whole time Staff Sergeant (SSG) Delmar G. Simpson had served as a drill sergeant (DS) at US Army Ordnance Center and School (USAOC&S). It was consistent with our objective of identifying potential victims and ensuring we provided all necessary support.

In an afternoon call with GEN William W. Hartzog, he confirmed that Sara Lister had been briefed and that CID would be briefing Secretary of the Army Togo G. West, Jr. the following Tuesday. We were getting a lot of exposure in the Pentagon. Visibility is good, exposure is bad!

I later learned that Sara Lister had asked for data on other installations, which indicated to me that the leaders in the Pentagon must have known APG was not the only installation with legal cases involving sexual misconduct by cadre with trainees.

MG James J. Cravens was the chief of staff for Training and Doctrine Command (TRADOC). His office called Captain (CPT) Jerry D. Stephens, my aide-de-camp, to say we needed to send information summarizing our situation to TRADOC headquarters so they could forward it to the inspector general (TIG) at the Office of the Department of the Army Inspector General (DAIG) by October 8. MG Guest had also asked for the same information so he could send it to GEN Hartzog in time for the four-star conference held right before the annual convention of the Association of the US Army in mid-October in Washington, DC.

CW3 Hayden and COL Buzz France came to my office and said the Army leadership had been briefed on incidents at installations other than APG/EA and that there was a growing concern about information expansion. For the rest of the day and evening, we focused on the information we would be providing to higher headquarters.

On Saturday, October 5, COL France came by and reported that he had learned in discussions with lawyers at TRADOC and the Pentagon that the focus is on the problem of sexual misconduct in the Army overall and not just APG. I felt good about this decision. At least we were starting with the right focus—Army-wide felonies. This focus would not last long.

At the morning update, CPT Stephens presented his initial work to compile the key data points on each subject and victim to see if we could find any common threads. Among other things, the data showed: all the subjects were either married, a geographic bachelor, or separated; all the subjects were over thirty; and all the subjects were black, except one, an instructor. At this stage of the investigation, the data also showed that the victims mirrored the ethnicity profile of the Army, which meant most of the victims were Caucasian.

I told the group we would continue to do what we were doing but that I wanted everyone to make absolutely sure we did everything by the book and make sure there was no bias in the investigative and legal processes. I knew that very morning that race would more than likely become an issue.

Later that morning, I sent what would become the first of 28 "vector reports" to an ever-increasing list of addressees at my higher headquarters (CASCOM, TRADOC, and Department of the Army). This first report consisted of 7 pages, and the last one I sent out on July 8, 1997, was almost 30 pages. I sent the reports every week up to January 14, 1997, when we went to a biweekly report. The reports were sent as attachments to emails to ensure timely submission to all addressees.

This first report included a cover page and a chronology of events, with a daily summary of actions with legal actions coded to serious incident report number to avoid inadvertent release of personal information. It also included a chart showing our three main vectors: one summarizing cadre–trainee incidents and data points for reference, a second identifying systemic causes, and a third covering preventive actions we were taking to preclude recurrence. The report informed recipients that we were reviewing females absent without leave (AWOL) during the past two years, suicide attempts in the past three years, Uniform Code of Military Justice actions in the past two years, and CID cases of former and current trainees for trends and for any incidents of sexual misconduct by cadre.

COL Raymond L. Rodon called to discuss the status of the 15-6 investigation. He informed me that a person in Sara Lister's office had called him directly the previous day looking for results of his investigation. Lister was a presidential appointee, and her office had a reputation of calling commanders directly on matters involving women.

On Sunday afternoon, COL Webb reported that a female trainee was allegedly being harassed and the chain of command was not doing anything about it. We asked CID and military police investigations to become involved to get to the bottom of the issue. My frustration that too many people just didn't get it was returning. I called Command Sergeant Major (CSM) Gerry Merrihew, the USAOC&S senior noncommissioned officer (NCO), and directed him to gather all master sergeants and above and emphasize the importance of the NCO chain of command working problems that were brought to them by soldiers.

CSM Merrihew and I reviewed the get-well plan with COL Webb, his deputy commander, and his CSM. We were making good progress, but still not at the pace I wanted. The major things we needed to do to ensure implementation were to standardize procedures among all units and to get the NCOs on board and working the issues.

During a lunch with officials from the surrounding communities, I had a discussion with Congressman Bob Ehrlich about the challenges we were working through. He asked me to keep J. P. Sholtes in his office informed. I said we would.

Later that evening, I received an email from MG Longhouser addressing what he could do to provide chaplain coverage at EA. He notified me that because the installation had lost three chaplains, he had to go to an area support system rather than have dedicated chaplains for each unit. EA would be covered by one of his remaining chaplains part-time and a contract chaplain two days a week. This was another example of cuts to help realize the peace dividend.

On the morning of October 8, I had a telephone conversation with BG Dan Doherty to cross-level information and coordinate the support his folks at APG required. This was the first of many direct conversations we would have over several months. I was confident BG Doherty and I shared the same objectives and the same commitment to doing the right thing. I thanked him for the super job CW3 Hayden and his CID team were doing and how much I appreciated CW3 Hayden's support for attending meetings and providing information while doing all of the investigations he had to do.

After the morning update in a stay-behind, I talked with CW3 Hayden and MAJ Gibson about the cases involving DSs that appeared to be in the Game.

I heard from MG Guest, my immediate superior officer, that GEN Hartzog was upset and concerned about the culture at training installations. This indicated to me GEN Hartzog knew what we had uncovered at USAOC&s was not an isolated incident.

I was subsequently informed that I would be required to present a briefing at the TRADOC commanders' conference at Fort Sill, Oklahoma, during a closed-door session on November 1. The TRADOC IG would speak for 10 minutes and cover sexual misconduct trends, and then I would speak for 10 minutes and cover actions taken, what agencies were brought in and when, and follow-on actions planned.

At the morning update on October 9, CW3 Hayden reported on the status of the investigation. In a stay-behind meeting, MAJ Gibson announced that CPT Robertson's Article 32 hearing would start later in the day. This hearing is analogous to a grand jury in the civilian legal system.

In the afternoon, I was at Fort Lee, and I called COL Jerry Luttrell at Redstone Arsenal. He confirmed they were checking procedures and looking into operations to make sure similar misconduct was not going on in any of the companies/detachments under his command. We had kept other elements of the Ord Corps posted on what was going on at APG/EA via our daily video teleconference for all organizations that were under my command or did Ord training. So far, problems existed at only 2 of 11 locations. Two, however was more than enough.

After getting a report on the status of COL Rodon's investigation, I was relatively confident we had identified the problems and were working to fix those under our control. I was also sure we were continuing to identify the problems others would have to fix.

The next morning, COL Cecily David reported on her analysis of female trainee visits to mental health services at the clinic during the period September 1995 to September 1996. The number of visits averaged about six per month. Two trainees had more than one appointment. This was consistent with the rest of the installation. The medical personnel had detected nothing out of the ordinary and no indications of sexual misconduct problems from the patients during this period.

In the afternoon, I received a call from Sara Lister's office notifying me that a team from her office, consisting of LTC Linda Thompson, LTC Lee, and Sergeant First Class Johnson, would be visiting us on Tuesday, October 15, to look at care of patients, CID operations, how we were keeping subjects away from victims, the counseling services available, and how AWOLs are being handled.

MAJ Pam Royalty came by to talk with me on October 11 about mental health care. She brought in several references for me to read and gave me a detailed briefing on sexual offender and sexual victim profiles with extensive literature to back up her points. This reinforced my belief that many of these young women needed adult leadership more than anything else.

I alerted her to the upcoming visit by LTC Thompson and her party.

As a follow-up to our previous conversation, MAJ Royalty informed me she had uncovered the fact that existing DSs were telling newly arriving DSs how to participate in the Game. I later learned the DSs in the Game thought they could pick out the female trainees who would voluntarily participate in the Game. They were not always right, and we would eventually learn of the problems that caused.

When GEN Hartzog called that afternoon, I gave him a complete update and told him I would talk with him on Sunday when I saw him at a gathering of former members of the 1st Infantry Division at Fort Meyer, Virginia. He said he had briefed the chief of staff of the Army and the other four-stars on the situation and concluded by saying I was "doing fine."

I made the decision that the whole crisis action team (CAT) would need to meet only once a day—in the morning—and that I would call

MAJ Gibson and CW3 Hayden in the evening for any updates I might need on investigations and/or legal actions. This was implemented on October 14.

On October 12, COL France gave me feedback on a meeting he had attended the previous Friday with the Army's top lawyers at the Pentagon. COL France reported that the focus of the meeting was concern over media interest and that the conversations were "grave and serious" and centered on the appearance of a black-male-on-white-female issue. It was obvious others had picked up that most of the subjects were black and most of the victims were white, just as I had surmised earlier.

Other topics discussed during the Pentagon meeting included the need for CID to review prior trainee abuse cases and the practice of DSs in basic training passing names of female trainees who were players to their DS buddies at advanced individual training sites.

The timing and location of a press release, the potential effect it could have on upcoming trials, plus the need to have the 15-6 investigation report available at the time of the release were also discussed.

COL France ended the conversation by informing me the Pentagon meeting concluded with a decision that the Army lawyers would write a letter for me to send to COL Rodon detailing additional areas for him to address in his investigation.

At the 1st Infantry Division get-together on Sunday at Fort Myer, Virginia, I provided an update to GEN Hartzog's executive officer. He forwarded it to the general later that evening.

Before introducing the team from Sara Lister's office to the CAT back at APG on October 15, I reminded everyone, primarily for our visitor's edification, that we were in the middle of ongoing criminal investigations and an interim AR 15-6 investigation. Therefore, we had to be careful in discussing sensitive information in an open forum.

At the end of the day, LTCs Thompson and Lee provided me their observations and recommendations. They agreed that some of the victims needed to be released from the Army and that we should consider using a Chapter 5 hardship discharge (chapter discharges are defined in Terms,

Definitions, and References), as this could be approved at the local level. The vast majority of their comments related to caring for the female trainees and was consistent with what we had already determined needed to be done. I felt we were on the right track.

We were, however, concerned about how the victims would get health care after they left active duty. This turned out to be a major issue that required a lot of work on our part.

CW3 Hayden reported that the 990 screening interviews would be completed by October 31, but that CID had decided to interview all 2,800 of the recent female trainees at USAOC&S. It appeared to me this direction must have come from BG Doherty as a result of the briefings done in the Pentagon. This decision made sense to me if we wanted to be sure we had identified all the victims. I concurred.

I took the morning update by phone on October 16 and started by reminding everyone that we should get our money's worth when visitors came. If we needed something they could provide, we should ask for it. I also asked that we look at where victims were living and working to be sure they were protected but not ostracized.

CW3 Hayden reported that CID had identified 3 additional subjects. We were now up to 12 cadre members alleged to have committed some form of sexual misconduct. This was a lot and too many, but I had to keep reminding myself that we had 1,145 cadre, which meant 12 out of 1,145 was just a little more than 1 percent. It was absolutely amazing the amount of work and trauma 1 percent could create.

While driving up Highway 195 from the airport in Austin to Fort Hood, Texas, I called BG Doherty to update him on the previous day's visit by the Lister team and to discuss the one trainee we really needed to get off of active duty for her own good. I gave BG Doherty the okay to investigate past commanders in our internal chain of command.

I then called BG Carl Freeman at Fort Bragg, and he confirmed he had approved the paperwork moving SSG Gamble back to APG for legal action.

I spent October 17–18 at Fort Hood with GEN Hartzog and the

other TRADOC commanders at the 4th Infantry Division, which was serving as the test bed for what the Army division would look like during Force XXI. The Army had developed a process to keep itself relevant into the 21st century. Force XXI was the process leading to Army After Next. This was the top priority project in TRADOC and was, therefore, my top mission priority.

Changing the logistics structure was a significant part of the process. It appeared to me the Army was reducing logistics personnel in order to increase personnel in the combat and combat support branches (e.g., infantry, armor, and artillery). Once again, we logisticians were going to be bill payers.

This was taking a great deal of my time, coordinating with commanders in the field and the Ord senior leaders to be sure the support structure could effectively support the force in the field. We were looking at our work with improved diagnostics, prognostics, mobile maintenance facilities, electronic technical manuals, tele-maintenance, automation, enhanced communications, and other initiatives to help offset antici-pated cuts in Ord personnel.

The weekend back at APG was relatively quiet. It appeared we were well on our way to getting to the bottom of the situation. CID was conducting interviews of current and former female trainees. We continued to analyze AWOLs, chapter actions, suicide attempts, and legal actions for trends. We had an action plan to address every problem identified by COL Rodon and by our own analysis. Folks were hard at work implementing the corrective actions.

5
★ ★

THE PROBLEM IS TAKEN PUBLIC

BACK AT ABERDEEN PROVING GROUND (APG), COLONEL (COL) Dennis M. Webb reported on the status of physically moving victims still at APG and Edgewood Arsenal (EA) under the command of the APG Garrison and into the Garrison's barracks. We were concerned because we were getting indications that some of the subjects (alleged perpetrators) were contacting their victims and asking them not to make statements.

I discussed with Command Sergeant Major (CSM) Gerry Merrihew how the noncommissioned officer (NCO) Corps was holding up. He thought the NCOs were doing better. One of our sergeants major, who was planning to retire, now wanted to stay on active duty. I took this as a good sign that at least one of our senior noncommissioned leaders wanted to stay and work the issues.

Major (MAJ) Susan S. Gibson informed me that Staff Sergeant (SSG) Wayne A. Gamble had been reported as absent without leave by his command at Fort Bragg. We needed to get him back to APG to process legal action on his alleged felonies.

I received a call from an action officer working for Lieutenant

General (LTG) Eric K. Shinseki, the Army's deputy chief of staff for operations (DCSOPS) in the Pentagon. The action officer was looking for information on trainee abuse cases so he could brief LTG Shinseki. We now had another staff section in the Pentagon calling directly to our headquarters. Satisfying all the requests for information from all three levels of command above us consumed a significant amount of our time.

Drill sergeants (DSs) Simpson and Gamble appeared to have the predominance of victims. Chief Warrant Officer 3 (CW3) Don Hayden reported at the October 25, 1996, morning update that Criminal Investigation Command (CID) would interview all 494 female trainees under Simpson's and Gamble's command while they were at APG/EA. SSG Simpson had arrived at APG in January 1995, and SSG Gamble had arrived in May 1994.

We had our first discussion of what the end state for all of this might look like. While I had the opportunity to work on other issues, folks like CW3 Hayden and MAJ Gibson had been doing essentially nothing but working sexual misconduct cases for almost two months. Just a few hours, every other day or so, of reading the statements of those involved was enough to drain anyone emotionally. I was amazed at how well these two great Americans were holding up.

At the update on Monday morning, October 28, CW3 Hayden confirmed CID was actively looking for all prior victims of felony-level offenses (rape, sodomy, indecent act, and indecent assault). He said there were problems with conducting interviews with potential victims who had participated in consensual sex in violation of our regulations and policies. Without granting immunity, some would not talk. This was an issue that would take months to sort through. As far as I was concerned, there was no such thing as "consensual" sex between a young trainee and an older DS who had almost complete control over her life.

On the morning of October 30, I flew from Fort Knox, Kentucky, to Fort Sill, Oklahoma, but I took the morning update by phone before I departed. SSG Gamble had been apprehended at his mother's house in South Carolina. The plan was for the military police from APG to drive down, pick him up, return him to APG, process him for pretrial confinement, and escort him back down to the Marine detention facility at Quantico, Virginia.

CW3 Hayden reported that CID would be getting assistance from other law enforcement agencies such as the FBI. He also said Sara Lister had told Brigadier General (BG) Daniel A. Doherty that "no sexual assaults would be left unnoticed." I wondered if this was throughout the Army or just in regard to our situation, but I let it go. I had enough to worry about without worrying about other commanders' problems.

Upon arriving at Fort Sill, I went to lunch with the leadership of the 19th Maintenance Battalion (Bn) and then visited Ordnance (Ord) leaders, soldiers, and students in training. At dinner that night, I gave Major General (MG) Ken Guest a copy of what I would be using as a presentation to the Training and Doctrine Command (TRADOC) leadership group on November 1. I also talked with the TRADOC staff judge advocate (SJA) and told him about the plans CID had to investigate students.

During our daily update, I learned SSG Gamble would be arriving back at APG the afternoon of October 31 and the detention hearing would be held forthwith. MAJ Gibson discussed the immunity issue for trainees who participated in consensual sex and said the SJA recommended blanket immunity.

I had meetings and discussions with MGs Guest and Cravens and COL Schemph (the senior lawyer on the TRADOC staff) about a closeout plan. I was concerned that without a defined end state, we would be in a constant react mode. Little did I know how prophetic that concern would turn out to be.

As the month of October came to an end, I was pleased with how much we had accomplished in not only working our way through investigations and legal actions but also reorganizing school operations and working to shape Ord support for Army XXI.

At the morning update on November 1, COL Johnnie L. Allen reported that the US magistrate denied the request to put SSG Gamble in pretrial confinement. He was now assigned to the APG Garrison, but living in the 61st Ord Brigade (Bde) area.

MAJ Gibson announced that MG John M. Longhouser had refused to sign the memorandum granting blanket immunity to female trainees. The way I understood it, MG Longhouser thought this would turn into a witch hunt all over the Army. This was not a good decision and would cause all of us many problems going forward.

At a breakfast meeting with Generals (GENs) Dennis J. Reimer and William W. Hartzog, I gave them an update on the situation. GEN Hartzog said he didn't think we could wait to put out a press release until we closed out all the actions. We discussed the timing. I was told LTG John A. Dubia, the director of the Army staff in the Pentagon, would be involved in the press release action.

At the TRADOC commandant's conference, my presentation followed the TRADOC inspector general's (IG's) presentation as planned. My briefing included my analysis of what had happened and what we were doing to take care of victims, how we were identifying alleged perpetrators, and what the emerging actions were that appeared necessary to preclude recurrence.

As I was looking out over the audience of my contemporaries, I saw two dominant expressions, at least as I perceived them, on their faces: (1) Thank God that's not me up there; and (2) I'd better make sure that stuff is not going on at my installation. The entry in my notebook read: "Brief went okay. I guess."

Before departing APG for Redstone Arsenal, Alabama, (RSA) on the morning of Monday, November 4, I signed a memorandum to COL Raymond L. Rodon. It detailed eight additional areas that needed further investigation before he submitted his final report. The primary focus was whether someone in the chain of command or a cadre contemporary knew of but did not report incidents of sexual misconduct.

Before the morning update, I called and talked with MG Cravens and BG Doherty. I got the distinct impression that pressure was building for a press release of some kind.

One of the first things I had done when all this started in September was buy Gregory L. Vistica's book, *Fall from Glory*, about the Navy's Tailhook sex scandal. I understood the importance of keeping higher headquarters informed and the need to put out a press release.

On November 5, MG Cravens called me at RSA to discuss additional points to put in a press release and to inform me I needed to be at a meeting in the Pentagon the next morning at 1000 in LTG Dubia's office to discuss the release.

When I called back to the team at APG to give them MG Cravens's guidance for the draft press release, MAJ Gibson expressed her concern that congressional notification was not being made in advance of the press release.

Captain Jerry Stephens scrambled to get us a flight back to APG. Jerry worked one of his usual miracles, and we got home late that night.

The team back at APG put together a memorandum for me to fax to LTG Dubia. It included the proposed draft of our press release and recommended questions and answers (Q&As).

In the cover memorandum, I documented CID's concerns that a press release would have a negative effect on the investigations and the Office of the Chief of Legislative Liaison's (OCLL's) concerns about notifying Congress before the release. I went on to say in the memo that while the legal community concurred in the release, they recommended a delay to avoid affecting legal actions.

LTG Jerry Bates, the inspector general (TIG) of Office of the Department of the Army Inspector General (DAIG), called me and recommended I take a woman with me to the meeting with LTG Dubia because Sara Lister would be there. I told him I didn't go anywhere without MAJ Gibson and had already planned to take her with me.

The next morning, I reviewed the draft press release and Q&As with the team and then had an abbreviated morning update before departing by van for the Pentagon with MAJ Gibson. We arrived at the Pentagon early, spent some time in the snack bar getting organized, and then headed to LTG Dubia's office. I asked LTG Dubia if MAJ Gibson could attend the meeting, and he said there was not enough room for her.

The attendees seated clockwise around a table were BG Gil Meyer, the Army public affairs officer (PAO) who had just taken the job a few days earlier; BG Doherty, the CID commander; MG Mike Nardotti, the judge advocate general of the Army; Tom Taylor, Office of the General Counsel; and me. A representative from LTG Vollrath's office and a representative for LTG Shinseki's office sat in side chairs.

LTG Dubia left to get Lister and escort her to the meeting. I had never met, seen, or talked with Sara Lister. Her entrance was not auspicious. As soon she stepped into the room, she said, "There's the problem. No women in the room."

I should have kept quiet but said, "My lawyer, Major Susan Gibson, is sitting outside since there was no room for her. Can I let her in?" The answer was no.

Lieutenant Colonel (LTC) Linda Thompson, who had visited APG a few days earlier, was with Lister. She stayed. There was plenty of room for MAJ Gibson.

Sara Lister ran the meeting. The first thing she asked was whether we thought we should put out a press release on APG. We went around the table in order: BG Meyer, yes; BG Doherty, no; MG Nardotti, no; Taylor, no; and me, no. Those of us who said no shared the common concern: the expected national media scrutiny would bring a potential negative effect and jeopardize the legal process.

Lister then said, "We have to act quickly, by the end of the week at the latest. OSD [Office of the Secretary of Defense] knows, and I'm afraid some SOB up there will give information to a friend in the press to curry favor. And then we'll be left holding the bag. The only thing positive about this is we found out about it and told people."

LTG Dubia agreed with Lister, so the majority was overruled, and she directed that a press release be published.

It was obvious that the decision to put out a press release had already been made by the secretary of the Army (SA) and that this meeting was a mere formality.

I discovered later how the folks in the OSD had learned of the situation at APG. A woman named Nancy in the public affairs office at OSD had been talking with an action officer in the Army public affairs office (PAO). When Nancy asked what Army PAO was working on, the action officer said the Army was working on rape cases at APG.

Sara Lister said she had read my proposed press release and that it was "woefully inadequate." I said I'd appreciate any help I could get because I was no expert at this kind of thing. She wanted to include the following in the release: zero tolerance was the standard because inappropriate behavior is not tolerated, the Army as an institution is people, and there was also misconduct at other installations.

Lister went on to say we needed to get help for MAJ Pam Royalty to handle mental health cases. I commented I could also use some assistance with medical resources because I was not the installation commander and those assets were not assigned to me. I'm not sure she understood that part.

Lister then announced she wanted an 800 number set up at APG for potential victims to call in and get help. I took this to mean the Army leadership expected only former trainees at APG/EA would be calling.

She also informed the group the secretary of the Army (SA) and the chief of staff of the Army (CSA) would be proactive to examine this problem with a study group and form an advisory board on sexual harassment. The CSA would also announce a chain-teaching program that stressed zero tolerance for sexual harassment.

Lister then said there would be two press releases: (1) GEN Hartzog and I would do a release and press conference from TRADOC headquarters. The TRADOC release and GEN Hartzog's statement were to cover the broader perspective and "not get down in the weeds." (2) GEN Reimer and Secretary Togo G. West would hold a press conference in the Pentagon, and the Army would put out a "green top," a Department of the Army official press release.

This was further confirmation that many decisions had already been made. There was no way Lister could have committed GEN Reimer and

Secretary West to such a course of action without their prior approval. I felt as if I were at the end of the whip as it was fanned through the air faster and faster.

LTG Dubia directed attendees to send a representative to BG Meyer's office to work on revising the draft press release.

Before heading down to the public affairs office, I called LTC Gabe Riesco and gave him the task of setting up a hotline and media center. With very little guidance (in reality, none) from me, LTC Riesco and the team found a building; arranged for an 800 number to service 13 lines for people to call, phone lines for the media, fax machines for our use and the media's, as well as tables and chairs; recruited volunteers to staff the phones; and appointed Captain (CPT) Paul Goodwin, my executive officer, to be the officer-in-charge of the combined media center and hotline location. The operation was up and running by 0900 the next morning.

We all sat around a table in BG Meyer's conference room and worked the press release for several hours. LTC Thompson and I were the only ones who had heard the guidance directly, which made it a little difficult.

One of the attendees at this meeting announced he had been at a recent meeting at OSD and a comment had been made to the effect that "heads would have to roll over this." For the first time, I wondered if mine was the one that would be rolling.

When we finished the revised draft, LTC Thompson took it to Sara Lister, who personally brought it back to tell us it wasn't what she wanted. We had not included a strong enough statement stressing zero tolerance. MAJ Gibson got into a rather heated discussion with Lister about the problems created by real or perceived command influence. MAJ Gibson knew her business and was very forceful, but to no avail.

Around 1730, COL Bob Gaylord (BG Meyer's deputy), MAJ Gibson, and I took the final draft press release back to LTG Dubia. The first paragraph read: "The Secretary of the Army and the Army Chief of Staff today expressed 'grave concern' about allegations of sexual misconduct and rape within the Army Training and Doctrine Command. They stated that sexual harassment is totally abhorrent to Army traditions and American values."

LTG Dubia approved the press release, and it was decided that I would fly to Fort Monroe first thing the next morning to prepare with GEN Hartzog and his staff for the TRADOC press conference that would be held at 1500.

Before departing the Pentagon late in the evening, MAJ Gibson and I went back to COL Gaylord's office to meet with BG Meyer and three media experts from his office who coached me on how to do a press conference because I had never done one. These ladies—Martha Rudd, J. C. Bean, and Bobbie Galford—were terrific. They had me practice a draft script and then shot practice questions at me. I cannot thank them enough.

The main points they stressed held me in good stead over the next few days and showed the importance of training senior leaders in advance on relations with the media: Be yourself; if you don't like the question, answer the question you wish you had been asked; don't get bogged down in nitty-gritty detail; develop your points and get them out as often as you can; and don't say, "Well, to be honest," because this indicates you may not have been honest in prior comments.

What a day. But as I would find out very soon, this would be just one of several unpleasant days.

Upon arrival at TRADOC headquarters the next morning, I met with GEN Hartzog in private. He told me the DAIG would be conducting an investigation, and the good news was they were not investigating me. I took this as a hint that I would eventually be scrutinized. I had learned that high-ranking folks don't mention something unless it has some significance. In this case, why mention investigating me if they weren't going to?

After our discussion, we were joined by the TRADOC PAO, Harvey Parrett. We were surprised to learn GEN Reimer would be making a statement at the Pentagon before our 1500 press conference. Despite some small glitches like this, one thing that worked well from the very beginning of this crisis was the constant dialogue among the PAOs at all levels of command.

I called John Porter in Senator Sarbanes's office, Julia Frifield in Senator Mikulski's office, and J. P. Sholtes in Representative Ehrlich's office to give them a heads-up about the press conference and the pending press releases. I called MG Morrie Boyd and LTC Scott C. Black to let them know I had notified our Maryland congressional delegation.

GEN Hartzog kicked off the press conference at 1500 and I followed. GEN Hartzog announced the DAIG would be conducting a TRADOC-wide investigation because he was "committed to looking at each and every TRADOC installation to ensure this is not a command-wide problem." Shortly after I began and right after I said, "This is the worst thing I have ever . . ." TRADOC CSM James McKinney, who was standing along a side wall by the podium, inadvertently pulled the plug on the feed to the Pentagon, where the Pentagon press corps was listening in. We did not know this had happened, and I kept on talking.

GEN Hartzog and I then returned to his office and conducted a teleconference with the Pentagon press corps. I assumed they must have been miffed by the abrupt termination of the feed from the formal press conference.

Once the local Aberdeen area, Baltimore, and Washington media learned there would be a press conference and press release, an installation-wide team began fielding calls at 1300 at the APG Media Coordination Center. The press wanted advance copies of the comments and releases. Ed Starnes told them they were not available, however, they could come to the Media Coordination Center to facilitate their coverage. Media reps began showing up around 1430. Representatives from at least 11 print and telecommunications media organizations conducted interviews at the center.

After the press conferences, calls were coming in to the hotline at a rate of about 100 per hour. Also at 1500, the Bn commanders at APG and EA began briefings on the situation for all staff, faculty, and students assigned to US Army Ordnance Center and School (USAOC&S) to let them know what was going on.

Each person was also given a small card that provided "rules of engagement for assisting the media" and directed people to contact Susan Gooch, our protocol officer, to arrange for media escorts. Susan was a real pro in the protocol business and worked unbelievable hours to ensure the press were properly escorted.

The six rules we printed on the card for all our military and civilian personnel were simple:

This is not a good news story. Don't make light of it.

If you are asked questions, it's your option to talk to the press.

1. If you decide to talk to the press, be friendly, be honest, be yourself, be serious (no jokes), and stick with what you know (just the facts).

2. Refer press to 278-0976, Bldg 5467 (Civilian Personnel Offices' Training Building) or call your chain of command.

3. Remember victims and subjects are entitled to their privacy.

4. Subjects are innocent unless proven guilty.

Upon returning to APG, I did an interview with Bradley Graham of the *Washington Post* from the phone in the van on the way from the airfield to my office. When we pulled up to my headquarters building, I found several large satellite trucks parked in the loop in front of the building. I did interviews with Channels 4 and 9 in Washington, DC, before heading home. I stuck with my main points, which were our objectives.

I was in bed asleep when BG Meyer called and said, "Boss, you did great today, but tomorrow is going to be really tough." I couldn't get back to sleep. I knew then and there I was no longer in control of the destiny of the USAOC&S.

6

★ ★

A MEDIA SPOTLIGHT SHINES
ON ABERDEEN

BRIGADIER GENERAL (BG) GIL MEYER WAS SPOT-ON. THE next morning, November 8, 1996, was a whirlwind.

I taped an interview with Bryant Gumbel for NBC at 0620. This was the first time I had ever done anything like that. I sat in a chair facing a camera. I could hear Gumbel, but I couldn't see him.

We had a conversation about the ongoing investigation, and then he asked me how I could have been oblivious to the misconduct. All I could think of was the guidance from the three ladies in the Army public affairs office (PAO) during my coaching session—answer the question you wish you had been asked. I said something like, "As I said, Bryant, we have three objectives here," and then proceeded to go over my key points. Gumbel pushed the issue in a couple more questions, and I kept repeating the three objectives. Because it was obvious I was not going to answer his question, he politely thanked me, and I was off the air.

From 0700 to 0930, I was live on ABC's *Good Morning America*, on Fox Channel 5 in Washington, DC, on *CBS This Morning*, and on CNN. I also did a telephone interview with Shawn Naylor of the *Army Times*.

At 0930, I was on NBC with Katie Couric. We had a pleasant chat, and then Katie brought on Susan G. Barnes, a noted attorney in the Denver area and a leader in the Women's Rights movement. We continued the conversation, and then Couric said, "You two stay on. I have another guest."

She introduced a woman who had been raped many years ago while she was in the Army. This lady was in obvious emotional distress. I could not see what was going on, but she was in tears, and it was really sad.

Katie then said, "General, what do you have to say about that?"

I said something like, "I think it's terrible. Someone needs to get this lady some help, for crying out loud."

I had a conversation with Susan Barnes about this interview a few months after I had departed Aberdeen Proving Ground (APG). She agreed I had been set up to come across as an uncaring male, but instead I came across as someone who really cared.

At 1000, I taped an interview with MSNBC, and then at 1100, I hosted a 30-minute live press conference on CNN. I gave a short introduction and fielded questions for about 20 minutes. When I got up to the podium, the room was packed. There were more than 30 news organizations from the United States, Japan, and Canada in attendance.

When I looked out over the audience, I wanted to say, "Hi. I've seen most of you on TV, and it's nice to see you in person," but I stuck to my script. At one point, I came close to losing my composure, and the flash bulbs were blinding. A little later during this press conference, I made the comment, "We need leaders in front of our troops, not lechers." This got quoted a lot and even appeared in a cartoon in the Colorado Springs *Gazette*.

After each interview, if at all possible, BG Meyer and his media expert, Colonel (COL) John A. Smith, critiqued my performance. They said I was getting better, and that was good enough for me. About half way through the morning, they said to stop calling this the "worst thing I had ever seen." A short time later, BG Meyer called back and said Secretary of the Army (SA) Togo G. West liked it, so I could go back to saying it.

I returned to my office around noon to catch up on work, and then at 1330, I presented a briefing to US Army Ordnance Center and School (USAOC&S) and APG installation personnel on the health of the Ordnance (Ord) Corps. I wanted to help them keep all events in perspective and make sure everyone understood that while we were focusing our main effort on resolving the sexual misconduct crisis, we still had a lot of other things to work on. I also wanted folks to understand it was only a small percentage of people who were doing bad things and capturing all the headlines.

According to the Training and Doctrine Command (TRADOC) PAO, the reason the press conference was conducted at Fort Monroe on November 7 was to keep the media away from APG. That didn't work so well for us and was just a harbinger of things to come because we were within easy driving distance of major media markets in Washington and New York as well as Baltimore and Philadelphia.

I started Saturday morning with a 0705 interview on NBC *Weekend Today*. That would be my last television appearance. SA West told BG Meyer he didn't want to see any more generals on television. I found out why a little later, but getting out of the spotlight was fine with me.

At the morning update, I learned COL Dan Quinn from Criminal Investigation Command (CID) and Lieutenant Colonel (LTC) Linda Thompson from Sara Lister's office had visited Capitol Hill the evening of November 8. They had briefed Senate Armed Services Committee (SASC) staffers Charlie Abell, P. T. Henry, and Peter Levine. The staffers were not happy because they had not received sufficient advance information before the public announcements—just as Major (MAJ) Susan Gibson told the folks in the Pentagon would happen.

COL Quinn and LTC Thompson could not answer all their questions, which were very probing. For example: "What's happening at other posts?" "Did anyone check Staff Sergeant (SSG) Simpson's prior assignments?" and "Why was the release done yesterday?" Abell commented that it looked like a two-star was in charge and that TRADOC and the Department of the Army (DA) were invisible. They also questioned whether I was the right person to be conducting the investigation. They questioned if I were truly disinterested: "Does Shadley have anything to lose depending on the outcome?"

It now made sense why SA West said no more generals on television. In days to come, SA West would indeed be the Army face on television. It also confirmed my feeling that I was no longer in control.

Public affairs activities occupied a lot of the discussion during the rest of the update. Our small headquarters was getting overwhelmed staffing the hotline and operating the Media Coordination Center. My staff had been reduced from 26 to 3 in the reorganization.

The operation at the Media Coordination Center was, however, receiving nothing but rave reviews from the media for its openness and the quality of support provided. This helped significantly to paint the Army in a favorable light compared to the Navy's handling of Tailhook. An editorial in the November 9 edition of the *Washington Post* was very favorable to the way we were being open and confronting the problem head on. It was followed the next day by a similarly favorable editorial in the *New York Times*.

The call center was a beehive of activity. As of November 9, we had received 1,528 calls. The 121 that were referred to CID for investigation were broken down into four categories: (1) fresh leads at APG; (2) fresh leads at other installations; (3) administrative referring to prior investigations; and (4) miscellaneous, not pertaining to any specific incident.

The majority of these calls had nothing to do with USAOC&S. We later divided the fresh leads at APG into USAOC&S and the rest of the installation. We needed more trained mental health workers and criminal investigators to work the phones. The people we had working the phones in many cases were just volunteers wanting to help out.

I sent Major General (MG) James J. Cravens, Jr. an email stating, "it's important to move the 800 number to echelons above the Ordnance School and APG." My assumption was the senior Army leadership had convinced themselves the problem was only at APG; therefore, they thought the only calls would be from those few cases we had missed during the CID interview process. A month-long battle began on November 9 to get the hotline moved to the TRADOC or DA level.

In the final accounting, when the hotline eventually moved to the DA level on December 12, only 4 percent of the calls referred to CID as potential felonies involved USAOC&S. In fact, there were more calls relating to the other organizations at APG than to the Ord School. We were, in effect, doing the work for the whole Army as well as the other services, with 96 percent of the CID referrals and potential felonies going to almost 60 other units, organizations, services, and installations around the world.

COL John Smith from Army PAO reported in on Saturday morning to serve as our full-time, on-site liaison with BG Meyer in the Pentagon. John quickly became an invaluable member of the team with his wealth of knowledge and experience in working with the media. His arrival was most timely; the media's thirst for data was becoming insatiable.

The next day, Sunday, was just another work day for the personnel in Building 5467, where the hotline staff was located. My wife, Ellie, and the other spouses began preparing food for the people working the hotline and in the Media Coordination Center so they would not have to go out for meals. I took my turn answering the phones for short periods. It was very enlightening, to say the least. Sunday morning became my time to work the phones for the next few weeks. Calls were now averaging more than 500 per day.

SA West appeared on CNN's *Late Edition* and talked about the number of cases and the number of cadre suspended in the ongoing investigation. He also said, "We live in a place where everyone makes a contribution . . . we have to pull from the largest pool of talent if we are going to have the best Army . . . women in the Armed Forces and the Army are doing their job . . . we will not go back."

Monday, November 11, Veterans Day, was a holiday for most but not the men and women making things happen at the Media Coordination Center and taking calls from all over the world.

The media frenzy continued. SA West was on Fox 5 News at 0700,

Fox Cable at 1200, and National Public Radio at 1400. General (GEN) Shalikashvili, chairman of the Joint Chiefs of Staff, was on the morning talk shows, and our own LTC Gabe Riesco was one of four guests on CNN's *Talk Back Live*. On Fox 5 News, SA West stated, "The purpose of the hotline is to give people a chance to tell us what's happening if they feel they don't want to confront it directly in a personal interview." It was good to hear the head of the Army say, in effect, it was the Army's hotline, not the APG hotline.

This was the fourth straight day of front page coverage in the *Washington Post*, but the first day of no coverage by the *New York Times* and the *Baltimore Sun*. The Army began looking at phasing out SA West and GEN Dennis J. Reimer on television and having Sergeant Major of the Army (SMA) Gene C. McKinney and subject matter experts take the lead for the Army.

CNN coverage began to shift to Fort Leonard Wood, Missouri, where three noncommissioned officers (NCOs) were facing courts-martial charges. None, however, were for rape or forcible sodomy.

Calls to the hotline remained heavy. The percentage of calls relating to organizations under my command decreased every day as callers mentioned other installations/organizations.

Tuesday was the first workday in the holiday-shortened week, and it resulted in the heaviest media day since day one. Visitors included Bradley Graham from the *Washington Post*, Greg Vistica from *Newsweek*, and Mark Thompson from *Time*. I was able to talk with Vistica, and I told him his book about Tailhook was of significant help to me in determining what had to be done in this crisis. The media thrust was shifting to in-depth analysis. For example: "Why was the chain of command surprised?" and "What should the Army do to correct problems?"

Fort Leonard Wood put out a press release on their three courts-martial cases. They emphasized these cases were being worked before the APG announcement and were not a result of the hotline.

This was also a heavy call day, with total calls now reaching almost

2,500 and the number of other installations mentioned growing. Fort Jackson, South Carolina, began to emerge as the most frequently referenced site for sexual misconduct.

At the morning update, we finally had a chance to tie things together and review our focus. Chief Warrant Officer 3 (CW3) Hayden reported that he had two agents working full-time on the hotline and that calls were heavy last night with Fort Campbell, Kentucky; Fort Hood, Texas; Fort Jackson, South Carolina; Fort Leonard Wood, Missouri; and installations in Germany and Korea most often mentioned.

I made the comment to the team that the media should back off APG, but that I did not think anyone in Washington was going to stand up and say sexual misconduct was an Army-wide problem; so, leave Aberdeen alone. I said, for whatever reason, our lot in life was now to walk point for the Army and take the heat so the Big "A" Army could work the problem and institute change that would benefit all soldiers. I had come to the conclusion we were on our own, and I made it clear we were going to do the right thing, work our objectives, and not care what happened to any of us personally. I was so proud of our team. Everyone embraced this approach and worked tirelessly over the weeks and months ahead to make it happen.

In my update call to Congressman Ehrlich, the race of the subjects came up, and this indicated to me that the topic was becoming of interest in Washington. No surprise there.

MG Cravens called and said I needed to be in the Pentagon the following morning at 0830 for a meeting to prepare Army briefers for two sessions on Capitol Hill.

I called MG Clair F. Gill, the commanding general at Fort Leonard Wood, Missouri, to cross-level information and gain any insights that might be of use to me at the session the next day in the Pentagon. MG Gill said they were not going back to interview all female trainees or screen past graduates. I thought this was a little strange, but it was not in my lane to decide what other commanders should or should not do.

The next morning, I left APG early to fly to the Pentagon to be at the

0830 meeting in the Office of the Chief of Legislative Liaison (OCLL) conference room. My understanding was that the team going to the Hill today would consist of John P. McLaurin, III, Office of the Assistant Secretary of the Army for Manpower and Reserve Affairs; Lieutenant General (LTG) Frederick V. Vollrath, deputy chief of staff for personnel, DA; MG Cravens, TRADOC chief of staff; and BG Dan Doherty, commander of CID. We reviewed the materials the team would use for their presentation. It was decided there would be no handouts provided to members of Congress.

The prep session lasted for about an hour, and then I went to BG Meyer's office. I may have been overly sensitive, but as I left the meeting, I had the distinct feeling some of those at the prep session thought I was carrying the plague and were reacting to me accordingly.

I strongly recommended to BG Meyer that we move the hotline. He talked about the clamor for data, and we decided to break out the definition of calls not referred to CID to help explain the delta in the data. The categories would be: crank calls, requests for information, make a comment (positive, negative, or neutral), and administrative. BG Meyer thought the attention would be shifting away from us and onto the Army overall with the SA's announcement that a panel would be established to provide a systemic look at the overall sexual misconduct issue. This phase would be orchestrated at the DA level. Unfortunately, this was not what happened or how things turned out.

I left the Pentagon and returned to APG. Johnnie Allen reported that Sara Lister's office had called looking for information on the number of calls about other installations. He also said the staff from ABC News *20/20* was wining and dining two female trainees in an attempt to get them on their program. We were also getting inquiries to the effect of, "Have any whites been investigated?"

MG Cravens called to give me a report on the morning briefing on the Hill. The staffers wanted more details and again questioned if I were the right level to be doing the Army Regulation (AR) 15-6 investigation.

Russ Childress met with and briefed local elected officials. He reported that we had their full support and that they recognized we were doing the right thing. Throughout this whole situation, the local community fully supported our efforts with regard to our three objectives.

There was a report in a daily Army PAO update that the secretary of defense announced a decision to look at sexual harassment throughout all the services. Every time I heard or read about an investigation into sexual harassment, I would think, that's fine—sexual harassment is a problem that must constantly be addressed so all people can feel safe and comfortable in the workplace. Sexual harassment often precedes felony sexual misconduct. But to me, the bigger problem was the felonies, such as rape and sexual assault. It was becoming obvious to me the military and civilian leadership in the Pentagon had a mindset that was akin to thinking that teaching everyone fiscal responsibility would prevent bank robberies.

On Thursday, November 14, one week after the TRADOC/DA press conferences, MG Morrie Boyd and MG Cravens called to give me a back-brief on the preceding afternoon's session on the Hill. Both reported the session went well but the issue was again raised about me being the right person to conduct the investigation. I took this to mean I was being alerted to an impending change of some sort.

At the 0730 update, Command Sergeant Major Gerry Merrihew announced SMA McKinney would be visiting us on Monday, November 18. SMA McKinney would end up being a major character in the unfolding drama of sexual misconduct in the military.

MG Joe Bolt, commander at Fort Jackson, held a 30-minute press conference and issued a press release announcing 30 cases of sexual misconduct at Fort Jackson over the past year. Three of the cases were unfounded, and 17 of the remaining 27 involved drill sergeants. Six of the 17 had been found guilty and were discharged, and the other 11 had also been found guilty but received punishment that did not include discharge.

In a subsequent phone conversation, MG Bolt told me, "Clair Gill [at Fort Leonard Wood], you, and me—we are all going to get fired over this." As I saw it, if they fired us, then GEN William W. Hartzog, who commanded the three of us and all the other TRADOC schools involved, would have to be fired. This could get real exciting real fast.

MG Cravens called me at the end of the day and asked for the numbers at other installations. I kept a spreadsheet listing all the 60 other units, organizations, services, and locations under which calls referred to CID for investigation of alleged felonies were logged. As of 1200 hours, November 13, the top 10 locations were the following, in descending order. (The number of incidents are shown in parenthesis, but they don't match exactly to calls because some calls had multiple incidents.)

1. Fort Jackson, South Carolina (44)

2. Fort Leonard Wood, Missouri (24)

3. Fort Dix, New Jersey (17)

4. Recruiting Command/multiple locations (17)

5. Fort Gordon, Georgia (13)

6. Fort Bragg, North Carolina (10)

7. Fort Bliss, Texas (9)

8. Fort Sam Houston, Texas (9)

9. Army/unknown location (9)

10. Germany (7)

It appeared the media was looking for a sensational story and that we at APG would stay the focus of media attention due to our proximity to the major media markets. Reporters in Washington and New York didn't seem too anxious to fly to Saint Louis and then drive hours to "Fort Lost in the Woods," as Fort Leonard Wood was sometimes referred. Jamie McIntyre of CNN would confirm this later with CW3 Hayden.

November 15 marked significant changes in what was going on both in the Army and at APG.

CW3 Hayden started the day with major news. He said the hotline phones would not be moved and he had a new requirement—to get a written statement from every caller within 48 hours of the call. CID operations around the world were now essentially shut down except for working sexual misconduct cases identified by us or callers to the hotline.

CW3 Hayden also reported he could no longer give me detailed data on calls relating to other installations. I now sensed someone or some group saw me as part of the problem and not part of the solution. Or that this whole mess was going to be painted as only an APG problem, and someone, or some persons, didn't want me to have data that showed otherwise.

MAJ Gibson faxed a memorandum to BG Meyer, expressing her concern about command influence. Defense counsel for some of the alleged perpetrators were raising the issue that their clients could not get a fair trial in the Army because of what senior Army leaders in the Pentagon were saying in the national media.

BG Doherty called, and we discussed the call data. The general breakout was okay, but no longer would the details by installation be provided unless a specific installation commander asked for that installation's call data. He said the same team that visited Capitol Hill on November 13 would return on November 18.

GEN Hartzog called around 0830 and said we at APG were "doing great work for our Army."

I sent MG Cravens an email, again strongly recommending the hotline be moved because it "has turned out to be the Army's Sexual Misconduct Hotline versus the USAOC&S Hotline." I went on to point out that "while this hotline is a most valuable resource to our Army and our soldiers, it is diverting my limited resources from our mission and from completing the AR 15-6 investigation." I noted that BG Doherty, BG Meyer, and MG John Longhouser were all being very supportive, but "we are also running this on a shoestring (for example, the spouses are cooking the meals for the workforce). Understand the USAF [United States Air Force] Hotline has 12 phones and computer stations. We are

13 stations with manual recording."

I talked with LTG John A. Dubia, who told me SA West had updated Dr. John P. White, the deputy secretary of defense, the night before. Dr. White wanted a point paper on the situation. He gave me the number to his fax machine so I could send him the information directly. I learned later Dr. White told SA West that West had better investigate this situation at his level—or Dr. White would at the Office of the Secretary of Defense level.

I received an email from MG Cravens that afternoon mentioning that the Army had not gotten information about sexual misconduct at TRADOC installations "through OCLL to the Defense oversight committees before major news stories [broke]." As a result, new reporting requirements were being instituted. This appeared to be another fallout from the November 13 briefing to the staffers.

GEN Hartzog called around 1700 and told me to stop my AR 15-6 investigation and send him all the materials. The DA inspector general or a TRADOC AR 15-6 officer would do the investigation. This confirmed my understanding of the verbal order Dr. White gave to SA West. This would also make the SASC staffers happy. I thought to myself, "If these folks don't trust me to sort out all of this at Aberdeen, then they should fire me."

I sent MG Guest an email advising him GEN Hartzog had stopped the investigation COL Ray Rodon was conducting for me. I went on to say, "I believe the Army wants to be sure that it appears we have a truly independent investigation—just like we asked on 15 Sep 96."

I assumed that whoever picked up the investigation would use the work we had already done. I feared we would lose 60 days of work if a new investigator started all over. This would later turn out to be the classic leadership challenge one works to avoid at all cost: I was responsible for correcting the problem, but I did not have the authority to make it happen.

Later in the day, I received a press release from Fort Leonard Wood announcing that four NCOs had been charged with improper personal relationships with trainees and other violations of the Uniform Code of Military Justice. As with the three previously reported cases of felonies, these cases happened before the hotline was set up.

As if we didn't have enough happening, the day ended with one of our trainee victims appearing on the television program *20/20*. It was indeed sensational. She reported 10 NCOs hit on her, most were married, one threatened to kill her, one raped her, and she didn't trust anyone anymore.

7

★ ★

THE PENTAGON TAKES CHARGE

O**N SATURDAY MORNING, NOVEMBER 16, I SIGNED AND** faxed a memorandum to Colonel (COL) Raymond L. Rodon directing him to cease his investigation and immediately ship all information he had collected to General (GEN) William W. Hartzog.

I spent a good portion of the day working the charts that would be used on Monday, November 18, to brief members and staffers of the House Women's Caucus in the morning and Senate staff members in the afternoon. COL Bryan Schempf, Training and Doctrine Command's (TRADOC's) staff judge advocate, called to discuss the charts. I told him COL Rodon had received the termination of investigation memorandum. The notes in my little green book about this call concluded with: "I get the feeling we are getting singled out."

I sent GEN Hartzog a lengthy email stressing my major concern about the question of "Who knew what, when, and what did they do about it?" I stressed that was the main focus of what I had wanted COL Rodon to do when he came back, and I recommended that whomever he chose to do the investigation should focus on that question. I felt we would never get to a conclusion unless we sorted out the accountability issue. This was another one that turned out to be very prophetic.

When I left Aberdeen Proving Ground (APG) eight months later, this issue had not been resolved. This was not a hard task to do. All one had to do was read the statements already collected by Criminal Investigation Command (CID). I eventually conducted my own investigation and turned it over to the TRADOC investigating officer, Lieutenant General (LTG) John E. Miller, in June 1997 before I left APG.

Sergeant Major of the Army (SMA) Gene C. McKinney and several members of his staff visited APG on November 18. Major (MAJ) Susan S. Gibson briefed the SMA on the cases. She mentioned that one of our lessons learned was the need for a code of conduct for drill sergeants (DSs), as it was obvious to us that several knew about the misconduct and either did not report their fellow DSs or covered it up. SMA McKinney responded to MAJ Gibson with: "When officers stop wearing backpacks into the Pentagon, I might consider an honor code for my NCOs."

This boggled my mind. The senior noncommissioned officer (NCO) in the Army was equating officers wearing backpacks (not part of the official Army uniform) to DSs committing sexual assault.

When I met with him, he lectured me on how to talk with soldiers, and he made it clear he knew he was really good at it.

The above irritants, combined with the fact that I was in the Media Coordination Center on November 10 when an anonymous complaint of sexual misconduct was called in against SMA McKinney, really set me off.

I was reaching for the phone to call GEN Dennis J. Reimer to respectfully ask him to keep his SMA away from APG, but decided I was not in a position to be pushy with the chief of staff of the Army (CSA). In hindsight, if I had called, I may have saved GEN Reimer some of the grief he would be forced to endure in less than 90 days.

Paul Boyce called at 1600 to give me a rundown on the briefings presented earlier in the day to members and staff of the House Women's Caucus and Senate staff members. Paul always took great notes, and his feedback tracked with the official minutes of the sessions published by Major General (MG) Morrie Boyd the next day.

Representative Patricia Schroeder expressed a concern other members of the Women's Caucus echoed about the possibility that some other members of the House would use these incidents to force a re-look at gender-integrated training. Representative Schroeder requested that GEN John M. Shalikashvili make a strong statement in support of gender-integrated training and the role of women in the Army.

The briefing team was scheduled to return to Congress the next day to brief the House National Security Committee (HNSC).

The media had been asking why we were transferring alleged victims to other installations. In our daily press release, we put in two paragraphs, which read:

> Some questions have arisen concerning the "transfer" of trainees from the Ordnance Center and School to other installations, specifically the transfer of individuals who have reported they were victims of abuse. These transfers are part of the normal reassignment process as the individuals complete their training.
>
> We have found that some trainees are reluctant to report abuse because they do not want to be held here [APG] pending completion of a case. Therefore it is important to continue the normal assignment process. Those soldiers who may be required to testify at hearings or courts-martial at a later date will be returned by the Army for that process.

In what was becoming almost a daily routine, I talked with Congressman Bob Ehrlich. He gave me some very sage advice on how to deal with certain members of Congress and the leanings (agendas) of key members.

Lieutenant Colonel (LTC) Gabriel Riesco passed along the good news that the hotline had now been officially renamed the "Department of the Army Hotline."

Our crisis action team (CAT) discovered that five black NCOs were making allegations that "the white man was putting the black man down." They and a sixth NCO were working on a letter to send to the media. The sixth NCO, who eventually had second thoughts, took the draft letter to CID and exposed their plot to use race as an issue to beat the charges against them. Two of the five names would surface later in another plot to get trainees to report they had been pressured to make complaints about blacks. This was just one of several instances of the "It worked for OJ, so it ought to work for us" strategy started by one of the alleged DS perpetrators and documented in official statements.

MG James J. Cravens called on November 19 to give me his assessment of the briefing and meeting with the HNSC. MG Cravens said Congressman Steve Buyer was the most vocal of the group. He was a Citadel graduate (as was GEN Hartzog) and had served in Desert Shield/Storm as a lawyer in the Judge Advocate General Corps. I received the official minutes from Office of the Chief of Legislative Liaison (OCLL) the next day: "The members asked the usual sets of questions focusing on time, accountability, what did the Army do with past historical data, why was the Army surprised, why no Hotline before, what policies and procedures went wrong, ombudsman for women, efficacy of gender integrated training, etc."

The minutes concluded with: "In a stay-behind that Mrs. Lister had with REPs Fowler and Harman (at their request), they advised that Chairman Spence had designated them to head a bipartisan delegation of members to Aberdeen to meet with the chain (of command), observe training and speak with soldiers in a 'town hall' type forum."

In a press release, Representative Fowler stated: "Since the system which is in place service-wide to prevent sexual misconduct appears to have broken down at Aberdeen and Fort Leonard Wood, Army officials are looking for other instances in which the system may not have performed properly."

Representative Spence's press release opened with: "I am very disturbed by the recent allegations of sexual assault and abuse at Aberdeen Proving Ground in Maryland and at other installations."

At least some in Congress sensed the problem was bigger than APG. Later, Representative Steve Buyer issued a press release saying he and Representative Fowler had been appointed to co-chair a delegation from the HNSC to visit APG in early December.

COL Buzz France called that evening to give additional feedback that the issue of women being trained in the same units as men had come up in the briefing to the HNSC, as had the issue of race. These two points were not in the official minutes, but would certainly be subjects of much debate and discussion in the months ahead.

At the morning update on November 20, I stated we needed to prepare for investigations of the race issue, to prepare to respond to gender-integrated training questions, and to get ready for LTG Miller's investigation.

I had a conversation with Congressman Bob Ehrlich, and he gave me a very useful insight about what to expect when the HNSC delegation visited. He also reported on a recent meeting he had had with Speaker of the House Newt Gingrich on the Army's sex scandal. He sent me a copy of a letter Speaker Gingrich sent to Representative Spence.

I spent a long time on the phone with MG Cravens covering a wide range of subjects, including moving the hotline, upcoming congressional visits, and his take on the briefings in which he had participated. MG Cravens said there were now four options in regard to the TRADOC-led investigation of sexual misconduct: (1) an Army regulation (AR) 15-6 investigation by LTG Miller with assistance from the Office of the Department of the Army Inspector General (DAIG); (2) an investigation led by DAIG only; (3) a DAIG investigation followed by a 15-6 by LTG Miller; or (4) an investigation by a TRADOC inspector general (IG) and DAIG joint team.

I had no sooner hung up with MG Cravens when COL Tom Leavitt from the DAIG called to say he and COL Tom Lopresti would visit us the following afternoon. They would do a reconnaissance of the area in preparation of their inspection and would depart on Friday, November 22. It was no surprise the DAIG would be involved in any of the above options.

I certainly agreed with the evolving Pentagon theme of "abuse of power." I agreed that the principal offenders did abuse their power. But to me, Playing the Game was also a major factor. I followed up my phone conversation with Brigadier General (BG) Daniel A. Doherty by faxing him a personal note that said: "Noticed on page 19 of today's 'Early Bird' [a daily news summary prepared by the Department of Defense (DOD) Current News Service] an article by Don Feder. He references a 1995 *Navy Times* article about 'the Game.' We've heard that expression here regarding drill sergeants getting sexual favors in return for preferential treatment. Know you probably were aware of this. Thought this was interesting though."

COLs Leavitt and Lopresti arrived as planned. They discussed where they fit in with the Secretary of the Army (SA) panel that was yet to be announced and the LTG Miller 15-6 investigation. I then escorted them to visit the Media Coordination Center and the hotline.

I gave COLs Leavitt and Lopresti an overview briefing on the Ordnance (Ord) Corps and the School and a very detailed briefing on the situation, which included all the details of the cases. I gave them hard copies of the current Vector's Report and a copy of COL Rodon's interim AR 15-6 report with all associated emails and memos. I promised I would provide them with the detailed timelines on our seven major cases, the get-well action plan listing the corrective actions we had identified and the status of actions to complete each, the daily CID status sheet, and a list of prior commanders and NCOs with addresses. I invited them to sit in our CAT meeting the next morning.

I started the morning update on November 22 by introducing COLs Lopresti and Leavitt and COL Luttrell and his leadership from the US Army Ordnance Missile and Munitions Center and School at Redstone Arsenal, Alabama, who were visiting to cross-level information and see firsthand how we were working the crisis. I made a point (primarily for our visitors) of reemphasizing our three main objectives and stressing that the number of charges would not equal the number of courts-martial. I went on to recommend to our DAIG visitors that the SA and the CSA tone down their rhetoric due to potential appearance of command influence. I also recommended that the director of the Army staff (LTG John A. Dubia) get our daily public affairs office update if he was not currently getting it.

At a Pentagon press conference at 1000, SA Togo G. West announced the membership and charter of and for the SA's Senior Review Panel on Sexual Harassment. He then released a copy of his November 20 Directive of Instruction to LTG Jerry Bates, the inspector general (TIG) of DAIG.

MG Richard S. (Steve) Siegfried was selected to chair the panel. MG Siegfried was a retired two-star who had been recalled to active duty for the specific purpose of chairing this panel. I had never met or heard of MG Siegfried, but he appeared to be a good pick. He previously served in the DAIG and as the commanding general at Fort Jackson, South Carolina, where, according to several of the hotline calls, a lot of sexual misconduct went on during 1991–94. With that background, he must have been SA West's choice as the most qualified person to head the panel. I assumed he would have had firsthand experience dealing with many felonies by DSs during his command tour at Fort Jackson.

Sara Lister was announced as the person who would exercise oversight of the panel and serve as SA West's liaison.

While at the media center later that day, I received a call from LTG Sam Ebbesen, an old friend serving as the deputy assistant secretary of defense for military personnel policy in the Office of the Secretary of Defense. He said: "Watch your backside, Shad Man. These asshole politicos [civilian political appointees] here in the building are out to get you, but we green-suiters [uniformed military] won't let it happen."

At the Saturday morning update, LTC Riesco related an incident that happened the day before when the male spouse of a deployed female soldier stopped by the media center. To document my growing concern that the sexual predators were part of gang-like structure, I sent an email on Sunday to MG Cravens, MG Ken Guest, and BG Doherty that read:

1. There was an article by Don Feder, of The Washington Times, that referenced "The Game" in a 1995 Navy Times article. We have heard that expression also. We sent you a copy of a tape called "Playing the Game" produced by a civilian firm. Not sure if there is a connection.

2. The male spouse of a deployed female soldier came to the Media Center on Friday and talked about: "GAM" which is short for "Game ala Military" Females in "GAM" are called "Freaks." We have also heard the phrase "locked in tight" when referring to females in the "Game."

3. Thought the TRADOC Response Team and/or IG may want to research this as part of your ongoing study/ investigation.

We received a call on the hotline from a former student at US Army Ordnance Center and School saying that "in 1992 many MOS 52D [generator mechanic] instructors were forcing female students to have sex with them. They told the students that they would recycle them [make the student retake the course] if they told anyone. They had the power to keep them there as long as they wanted."

This indicated to me—and it was not the only indication—that this pattern of sexual misconduct by DSs and instructors had been going on since at least 1992 at APG. I would later learn this type of misconduct had been going on at other training bases long before that. The Game was on and being played at several locations in the Army.

On Monday, November 25, Senator Mikulski's office put out a media advisory saying she was going to meet with me and trainees at APG on November 26 and then have a press availability at 1215.

Chief Warrant Officer 3 (CW3) Don Hayden reported they had

uncovered a conspiracy among at least three DSs in the 16th Ord Battalion. Over the next several weeks, this would play out as a significant cause of additional work created by outright lies that these three had been discriminated against. Unfortunately, only after a lot of damage had been done did one DS finally agree to turn informer and come clean on what they had done.

CW3 Hayden also reported they were continuing to receive numerous complaints about Fort Jackson in the 1991–94 time frame.

I have an entry in my notes for the morning of November 26: "Note for the record. Concerns about Army study of sexual harassment/abuse/ misconduct. SMA [McKinney]: attitude, integrity, and complaint to hotline about his office. Team Chief [Siegfried]: Don't know him, but many cases referred to CID about cases in 1991–1994 time frame at Fort Jackson."

Senator Mikulski arrived on November 26, accompanied by Julia Frifield and Claire Hassett of her staff and LTC Tom Hawley from OCLL. LTC Hawley's minutes of the meeting and Claire Hassett's press release afterward were both indicative of a "good visit." It did turn out okay, but my notes indicate I had a difficult time.

The senator entered my office with the other folks, and I offered her a seat in one of the two easy chairs I had in the office. She declined because she said her feet would not touch the floor. So we all sat around the conference room table in my office.

Before we had even started, she said she wanted me to see a letter she had sent to GEN Shalikashvili, chairman of the Joint Chiefs of Staff, before she left Washington that morning. The letter asked him to set up a task force at the Joint Chiefs level to look into the matter of sexual misconduct. Two sentences caught my eye: "Either the base commanders were out of touch, or they knew and took no action. The role of base commanders must be examined." I thought, "Great. She thinks I should be fired, and we haven't even started the visit."

I then asked her what she would like to discuss. She asked where her briefing was. I replied that her staff said she didn't want a briefing, that she just wanted to talk with me. She said she wanted a briefing. I turned to MAJ Gibson and said, "Susan, give the senator a briefing." MAJ Gibson did her usual thing and thoroughly impressed the senator with her understanding and depth of knowledge. This went on for a while, and then the senator said she had heard enough and wanted to talk with trainees. We had 15 female trainees available for her to meet in a private room. The trainees represented a cross-section of ethnicity and were not ringers. My guidance before all these sensing-session type meetings was to select a cross-section and let the trainees talk. If something bad came up, we would work the issue at that time.

After about an hour, the senator opened the door of my office and proclaimed, "General, what's all the fuss about? With the exception of one soldier who needs a medical appointment, you don't have a problem."

I should have kept quiet, but I figured, what the heck. So I replied, "I don't know, senator. Why don't you tell me? You're the one who wrote the letter to the Chairman saying I should be fired." She smiled, and we sat down and had a nice chat.

I did not go with the senator and her party to the media center for the visit and the press availability, but I understand she was impressed with what we were doing with the hotline. She told her staff they should have gone there first.

It was obvious she considered sexual misconduct an Army/Department of Defense problem and not just one isolated to APG. I concluded the day with a genuine respect for Senator Mikulski's drive, intelligence, and willingness to accept new information and modify her assessment of the overall situation.

I called GEN Hartzog and gave him my take on the Senator Mikulski visit. I also recommended that TRADOC needed to develop procedures for handling old cases going back several years. We had one case in 1985 and another in 1990. I again stressed that we needed to get the hotline moved off of APG as soon as possible.

On November 27, TRADOC headquarters sent out a 7-page set of instructions on how to fill out and submit its new TRADOC Sexual

Misconduct Report on a weekly basis. This appeared to be TRADOC's attempt to overcome the criticism that the DOD did not comply with the 1989 congressional order to standardize sexual misconduct reporting. DOD asserted the order was never implemented because the DOD did not receive the corresponding funds to implement it. That's why we had no reference data when we started working the problem almost a year earlier.

The good news was that we were beginning to see some positive benefits from our coming forward with a problem.

The president of the Harford County (Maryland) Chapter of the NAACP called the media center on November 27 and requested she be allowed to sit in on any meetings we had with members of Congress. This would prove to be a precursor for extensive dealings with the local chapter of the NAACP.

On Thanksgiving, I visited the dining facility at Edgewood Arsenal and talked with soldiers enjoying the huge meal that is traditional in the Army. Ellie and I then ate at the dining facility at APG. We swung by the media center, and all was quiet. Ellie and the other spouses took treats over the holiday weekend to the men and women staffing the hotline.

Gilbert A. Lewthwaite and Joanna Daemmrich wrote a lengthy article in the *Baltimore Sun* titled, "Fort Jackson recruits recount sexual harassment—women tell about rogue drill sergeants, obscene cadences at Army Training Center." The article stated: "Between October 1994 and November 1996, 87 allegations of rape or other sexual misconduct were lodged at Fort Jackson. Of the total, 51 involved military trainees as victims, and 33 validated charges involved drill sergeants. More recently, at least 85 of the accusations made on a special crisis hot line at Aberdeen have been referred back to Fort Jackson for criminal investigation."

What a month November 1996 turned out to be. Unfortunately, we didn't know it then, but we were in fact still closer to the beginning than the end.

My problem with a few folks who worked agendas such as race and women in the military was that in too many instances, the agenda took precedent over the individual. Some people would not hesitate to throw someone under a bus if they thought it would further their agenda. I would eventually get a view from under more than one bus.

8

★ ★

THE RACE ISSUE SURFACES

A T THE MORNING UPDATE OF DECEMBER 2, I ALERTED everyone to an impending visit by representatives from the National Association for the Advancement of Colored People (NAACP). I thought we would be in pretty good shape for this visit. We had surmised early on that due to the alleged offenders being predominately black, race would become an issue. Chief Warrant Officer 3 (CW3) Don Hayden and his Criminal Investigation Command (CID) team had worked very hard to ensure we had a representative group of agents by gender and ethnicity and that all the interviews were done correctly. Lieutenant Colonel (LTC) Gabriel Riesco advised that a team from the US Army Personnel Command in Alexandria, Virginia, would be at Aberdeen Proving Ground (APG) the next day to work the transition of the hotline from APG to their location—great news.

At the morning update on December 4, I recommended to LTC Riesco that we shoot for December 15 as the date for the hotline transfer. Johnnie Allen reported that the House National Security Committee delegation was still set for December 11, but that the NAACP had requested postponing their December 5 visit to a later date.

81

General (GEN) William W. Hartzog called to say he was hearing that some of our victims were having problems. We had six victims still at APG. Three had major emotional problems the mental health folks were working through as well as looking at administrative solutions for their problems. One of these three would be a major concern for weeks to come. She was exactly the type of person the sexual predators preyed upon and was a perfect example of why young soldiers such as her needed a leader and not a lecher for a drill sergeant (DS).

I had met Greg Vistica a while back and told him his book, *Fall from Glory*, about the Navy's Tailhook sex scandal was very useful in helping us plan how to deal with the situation we found ourselves in. COL John A. Smith got him to autograph my copy when Vistica made a return visit. Vistica's note read: "To Major General Robert Shadley, You've got your work cut out for you, but I'm positive your honesty and dedication will solve this cultural problem the Army is dealing quite well with."

The next day, I was at another meeting at Fort Lee and stayed apprised of the legal actions via regular calls back with Ed Starnes, who was doing his usual superb job of note taking and working with the media. There were 25 media organizations represented in the court room as Captain (CPT) Derrick A. Robertson and Staff Sergeants (SSGs) Delmar G. Simpson and Nathanial Beach were arraigned.

The good news was, we were making progress. The bad news was, we would have media interest at APG, as a minimum, for another three months even if nothing else happened.

We received good, but somewhat ironic, news on December 9. The results of the Defense Equal Opportunity Management Institute (DEOMI) survey Colonel (COL) Dennis M. Webb had requested on July 24 finally arrived in the mail. The survey showed the overall equal opportunity (EO) climate for the 61st Ordnance (Ord) Brigade (Bde)—including the 16th Ord Battalion (Bn) and the 143rd Ord Bn, both military and civilian—was rated "Above Average."

The overall rating for the military was "Above Average," with each of the 11 subcategory numerical scores being good and the narrative for each indicating a positive EO climate with a small chance of sexual misconduct occurring.

The overall rating for civilians was also "Above Average," and the individual subcategory scores were essentially the same as for the military.

I thought to myself, "If we are above average, what is going on in below-average organizations?" I lamented, "So much for the accuracy of the DEOMI survey," which was, to our knowledge, the only tool available at the time for commanders to use to assess their commands.

I sent an email to my higher headquarters and key folks in the Pentagon alerting them that we would retain the hotline at APG through the CODEL visit on December 11 so CODEL could see the operation per their request. I also advised that our last daily update on the calls would be on December 12.

I received a nice reply from Major General (MG) Morrie Boyd: "Thanks my friend . . . please know that you have a lot of your buds impressed with how you have been handling this . . . good luck on Wed [with CODEL], Morrie."

Congressman Bob Ehrlich called to give me some additional tips on dealing with our congressional visitors and provide background information on the key members. He said the two main issues were: (1) how could this happen? and (2) what proactive steps are being taken to preclude recurrence?

After the morning update on December 10, I met with COL Tom Leavitt from the Office of the Department of the Army Inspector General (DAIG). He advised that both Lieutenant General (LTG) Bates, the inspector general (TIG), and MG Larry R. Jordan, the deputy inspector general (DTIG), would be visiting APG in the near future. We discussed some other administrative details, and then he gave me a list of initial observations from his team: (1) APG garrison orientation of newly arrived trainees was a problem—APG did not have an effective in/out processing center; (2) chain of command involvement was missing—hands-off approach by leaders; (3) the training that is going on is good;

(4) we are under-resourced (people and funds); and (5) we needed additional assets.

I later talked with Russ Childress about the hands-off comment. He thought that must have been brought up by people talking about how my predecessor, MG Jim Monroe, had operated. Russ said MG Monroe had told APG commanders and staff during his command tour that he had big issues to work and didn't want to be bothered by details.

The major event scheduled for December 11 was the CODEL visit. I had an office call with MG Jordan, DTIG, at 0700. MG Jordan stayed for the daily 0730 update, which focused on two female trainees with significant emotional problems. MG Jordan commented at the end that we were "doing good."

I was standing tall on the ramp at Phillips Army Air Field when CODEL's aircraft landed at 0845. We had a bus to drive us over to MG John M. Longhouser's headquarters in Building 314 so we could use his conference room, which had state-of-the-art audiovisual equipment and enough room to seat the whole delegation at the head table facing the platform. Representatives Molinari and Ehrlich drove separately and joined the group a little later. LTCs Tom Hawley and Scott C. Black from the Office of the Chief of Legislative Liaison (OCLL) also flew in with the other 11 members and delegates.

The whole day seemed to go well. I called MG Boyd's office immediately upon their departure to give him a verbal report and then sent the following via email:

1. The purpose of this email is to give you a quick AAR [after action review] on our visit today by . . . 13 Members of Congress:

2. The group's itinerary included a briefing by me and my brigade commander, COL Webb; three focus group discussions with trainees (16 in each group); focus group discussion with 15 drill sergeants; and a focus group discussion with company commanders and first sergeants (total of 21). The Congressional Delegation

(CODEL) then had a working lunch with our victim support team in the dining facility. Just prior to departure, the group held a press session at the airfield.

3. The CODEL's assessment was:

    a. Very appreciative of us being so open and honest.

    b. Think we are doing the right things.

    c. Believe there are larger issues other than just APG—one being that the Army appears to be under-resourced in certain areas. Several members of the Caucus on Women's Affairs continued to emphasize that we needed an ombudsman for women's issues.

    d. Media appeared to be miffed at the press conference because the majority of the comments by the members were positive.

4. Will defer to OCLL representatives in attendance for their professional assessment of the visit.

MG Boyd sent me an email at 1644 that read: "Thanx . . . hotwash from Hawley/Black says same . . . Details being worked as we speak . . . well done to all at Aberdeen. . . ."

I did not attend any event except the in-briefing, but I stood outside the hanger door and listened to the press conference, which included 42 media representatives from 24 news agencies.

Since Representative Molinari had driven directly to Building 314, I gave her a ride back to her car as the others flew back to Washington. I was really impressed with her questions and her depth of knowledge and understanding of the big picture. I sensed she thought I was okay and that we were doing the right things.

Our officers, noncommissioned officers (NCOs), soldiers, and civilians showed what the vast majority of Ord Corps personnel were like. I received the following note from LTC Hawley the next day: "Sir, you requested hard copy bios with pictures of the members who visited APG yesterday. They are attached. By the way, all comments on the ride back were positive. You and your soldiers made a great impression on some members who had been skeptics. Johnnie Allen did a superb job in coordinating a very tight schedule handed him by the delegation. Thanks for all your help."

I was really proud of our team—COL Webb, LTC Allen, LTC Riesco, and MAJ Gibson in particular—for making all the moving parts of the CODEL visit fit together. These great Americans spent countless hours making things happen. I will be eternally grateful to them, as well as to many others.

The big news story for December 12 was an article in the *New York Times* reporting there were 25 subjects and 50 victims at the US Army Ordnance Center and School (USAOC&S). I spent most of the day working numbers with various staff sections at the headquarters of Training and Doctrine Command, CID, and Department of the Army.

This precipitated a rash of calls from within the military asking about the numbers and requesting information. Representative Molinari also called, and I gave her the data I had on alleged subjects and victims. We discussed the importance of staying away from numbers to avoid jeopardizing cases. As we were seeing in ongoing legal proceedings, not all charges were substantiated and the numbers of victims and subjects were changing. I appreciated Representative Molinari calling me directly. She was a straight shooter, as Bob Ehrlich had said she would be.

We topped off the day by putting out our last media update on the hotline. We informed everyone that as of noon on December 12, the hotline was now being operated by the US Total Army Personnel Command in Alexandria, Virginia.

In an official media update dated December 12, we provided a detailed accounting of calls received during the 26 days the USAOC&S ran the hotline for the Army. The key data were that of the 6,250 total calls, 850 had been referred to CID for investigation of possible sexual misconduct. Of the 850 hotline calls referred to CID, only 34 calls, or 4 percent, involved units under my command at APG. The exact breakout of the 850 referred calls was 34 calls for units under my command, 87 calls for other units at APG, and 729 calls for other locations.

The percentage of calls relating to my command would drop even further over the coming weeks when the hotline was operating under new management.

As with the congressional visit the day before, I was very proud of the CID, Garrison, and USAOC&S team (supported by a group of caring spouses and family members) that took on the hotline and media center mission on short notice with no guidance or concept of operation and no end state. Our team received almost universally positive comments about how it dealt with every caller and provided support to over 200 media sources over a 26-day period in the national spotlight.

BG Gil Meyer advised me that Secretary Togo G. West was worried about who had gotten the numbers of victims and subjects and released them to the media. He said to expect more news articles about the numbers of victims and subjects. We did not release the numbers. If the Army staff hadn't, then I could only guess who did—the Office of the Secretary of Defense?

Near the end of the day, I called John Porter in Senator Sarbanes's office. He let me know the senator had met with Secretary West the previous Wednesday (December 6) and the racial issue was discussed. This confirmed my understanding that Secretary West knew race would eventually become an issue.

I was in Alexandria, Virginia, the first thing Monday to meet with GEN Johnnie E. Wilson, the commanding general of the 80,000-person US Army Materiel Command (AMC). He was also the senior ranking Ord officer on active duty, and I coordinated all actions relating to the Ord Corps with him, as I had done with his predecessor. In our conversation, we did talk about several scandal-related items. I briefed him on the demographics of the victims and subjects and told him race was becoming a major issue. GEN Wilson was not only a four-star general and the senior Ord officer on active duty, but he was also black. I had known him since 1980 and worked right down the hall from him when I was the executive officer and he was the chief of staff for GEN Jimmy D. Ross at AMC in 1992–93.

We discussed the need for a code of conduct for DSs and more emphasis on values in training. I also solicited his support to build an in-and-out processing center at APG as recommended by the DAIG team. He would have to make that happen, as funding requests for projects at APG would have to come from him and not GEN Hartzog.

LTG Bates sat in our morning update, which had been delayed because of my trip to see GEN Wilson. He commented that many of our procedures were based on tribal wisdom and not standing operating procedures and this led to the variance in how the two Bns were operating. We had already identified that issue and were working corrective actions; therefore, this observation was not a surprise, but it did confirm our internal assessment.

People were beginning to tell us things we had already figured out. More than anything else, we needed space and time to bring all the legal actions to conclusion and correct the discrepancies within our control that had contributed to the problems predators in gangs created. This issue was superimposed on the same problems every other installation had with regard to male–female relationships and associated sexual harassment and/or felony sexual misconduct. All this was confounded with a severely flawed Combined Arms Support Command reorganization plan and implementation.

Congressman Ehrlich's weekly column for the December 16 edition of his publication *Ehrlich in Washington* was reprinted by *Army Link News*. Several paragraphs summarized the Congressman's impression of what we were doing in regard to our situation:

> Last week, I participated in a bipartisan congressional visit to the U. S. Army Ordnance Center and School at Aberdeen Proving Ground, Md. As APG's representative in

Congress, I played host to 12 of my congressional colleagues, including three fellow Marylanders.

I have visited APG approximately a dozen times since I was elected in 1994. I always look forward to the visits. . . . More than 12,000 civilian and military employees work on the post. They perform their duties with dignity and honor.

Regrettably, the alleged activities of a few individuals have temporarily overshadowed the work these fine men and women perform for our country. As you are aware, serious allegations of sexual misconduct and rape at the Ordnance Center and School have drawn national attention. Concern over these allegations brought my colleagues and me to Aberdeen last week.

I am deeply disturbed by these allegations. . . . Reports of sexual misconduct are completely unacceptable and must be dealt with swiftly and severely.

I have been in contact with Maj. Gen. Robert Shadley, commander of the U. S. Army Ordnance Center and School, on a daily basis since these allegations became public. I am pleased by the quick, proactive steps taken by the Army in response to the problems reported at Aberdeen and at other military installations.

Moreover, the Army and Congress are determining the extent to which policies and procedures contributed to this situation. The Army has pledged to implement all necessary and appropriate reforms.

On December 17, MG Cravens sent an email to all TRADOC commanders addressing the issue of holdovers, which resulted from poor data in the personnel system, delays in completing criminal background checks, delays in finalizing security clearance investigations, and delays in the medical system. This was welcome news that our higher headquarters was working to help us with these administrative problems that kept soldiers from going to their next assignments and being productive in the military occupational specialty for which we had just trained them.

Late in the day, I practiced my briefing for the NAACP delegation with the Bde commander, the Bde deputy commander, the Bde adjutant, Russ Childress, Johnnie Allen, and Major (MAJ) Susan Gibson. When I finished, I asked the group what they thought. Going around the table, everyone thought it was great, except MAJ Mary Joe Clark, the Bde deputy commander. I asked MAJ Clark what was wrong with it, and she said I sounded like a racist. I was taken aback and said, "What do you mean, I sound like a racist?"

MAJ Clark, who was black, said words to the effect, "I know you are not a racist, but that's how my black ears heard what you said."

I have used that comment hundreds of times since that day as a good example of why it is so important to have gender and ethnicity diversity in meetings when decisions are being made by a leadership team. Based on many factors, not everyone hears the same thing or hears it the same way the speaker intended. It was also important for all groups to see they have a representative at the table when decisions affecting them are made.

Right after MAJ Clark's comment, the brigade adjutant commented, "Quite frankly, General Shadley, we in the African-American community have a problem with our males, but you will never hear us talk about it in public."

I had been worried about the NAACP visit on December 19 before we started the prep session, but now I was really worried. But I figured that if I presented the racial breakout for the victims and subjects compared to the Ord Corps and the Army data, it would make the case that our objective was to find victims and not target blacks to be subjects. My analytical mind was convinced the data would prove our point.

During the morning updates when CW3 Hayden presented basic data on new subjects, I said a silent, simple prayer to myself: "God, please let them be white." But it was what it was, and I could not change it. In one private conversation with CW3 Hayden, I remember saying something like, "Don, I can't believe there is not at least one white drill sergeant who's doing this crap." CW3 Hayden was very professional and continued to go to where the alleged victims pointed his investigators.

I called GEN Hartzog to update him on our upcoming meeting with the NAACP and the results of the DAIG out-brief. GEN Hartzog said LTG Bates had told him the teams had not found anything we at APG did not already know. I felt good about that but wanted to get on with the corrective actions so we could get out of the spotlight.

COL Smith called and said Eric Schmitt from the *New York Times* had asked GEN Ronald H. Griffith, vice chief of staff of the Army, if they were going to fire me. This was becoming almost a weekly cycle of some people saying they heard I was going to be fired and then others saying I was really doing well.

My executive officer, CPT Paul Goodwin, reminded me that when he had asked me if I were worried about getting fired I had replied: "No. I'm worried they're going to make me stay here until I get it right." It was probably not the right thing to say, but I figured I'd be okay as long as I kept a sense of humor.

I spent most of the day on December 19 preparing for the NAACP visit that evening and finalizing the press release covering the number of charges against SSG Simpson.

That afternoon, I met with the defense counsels for two alleged perpetrators to tape my testimony regarding command influence. In addition to the legal clerk doing the recording, MAJ Gibson and CPT Jerry Stephens sat in. The defense counsels were concerned about my comments, "the worst thing I have ever seen" and "no such thing as consensual sex between a cadre member and a trainee." They covered the sequence of events, what I knew when, and my conversations with commanders. I made the point that I was not the general court-martial convening authority. That authority rested with MG Longhouser, the installation commander.

The meeting with the NAACP started at 1820. The NAACP delegation consisted of Leroy Warren, national board member (Criminal

Justice Committee); Janice Grant, president, Harford County Chapter; Reverend C. E. Hunter, president, Baltimore County Chapter; and Dr. Bernetha George, Baltimore, Maryland.

Those in attendance from APG in addition to me were COL Buzz France, COL Webb, COL Cecily David, LTC Allen, LTC Riesco, MAJ Clark, MAJ Gibson, Ms. Bruce, Command Sergeant Major (CSM) Gerry Merrihew, and CSM William Miller.

I started with an overview briefing of the school and the situation. I included a chart that gave the ethnicity profile of the victims and alleged perpetrators compared to the overall populations of the Army, the Ord Corps, and the Ord school at APG.

I tried to make the point that the percentage of black cadre subjects was high because we had a high percentage of black sergeants in the overall population of our cadre, especially DSs. I also pointed out the victim profile was fairly close to the overall student profile. My point was that the alleged perpetrators were "equal opportunity" sex offenders—they didn't care what race the females were they tried to have sex with.

COL France commented that as a father, he would want to make sure his daughter was in a safe environment in the Army. Mr. Warren became very angry and lashed out at COL France: "See—you all are just worried about your white daughters around black men." The tone then became that the delegation was convinced we were racist and charging the black DSs because of their race. Among my notes on the comments by Warren were, whites are doing the same thing and not being punished; blacks are being selected for tokenism, and it's a witch hunt focused on blacks; people are under pressure because of white daughters; cases are really love affairs gone bad; and some females are lying to the press.

Janice Grant then related her story, which would later be documented in the March 17, 1997, edition of the *Baltimore Sun*. In 1963, she married a military policeman who was stationed at APG. They went to France on their honeymoon. When they returned to the United States, they decided to go to the Martin Luther King address at the Lincoln Memorial in Washington instead of returning to APG. As a result of not coming back per her husband's authorized leave schedule, her husband received a month of extra duty as punishment. She said her husband had to fight racism in the military, and she felt for the DSs who were also victims of racism.

She then went on to relate an allegation that SSG Wayne A. Gamble was poorly treated when the military police (MP) picked him up in South Carolina for being absent without leave and returned him to APG. Specifically, Mrs. Grant said the MPs would not stop the car to let SSG Gamble go to the bathroom, and as a result, he had to urinate in his pants. Then upon arriving at APG, he was paraded around the gymnasium (before using the shower facility) for all to see, including his accusers. We had no knowledge whether such events had actually occurred.

It was readily apparent this was a no-win situation for me. I told the delegation we were absolutely committed to fairness and equal treatment for all. I stressed our first objective was to find victims. The victims identified the alleged subjects to the CID investigators. I said we were not out looking for black subjects because it did none of us any good to have all the great black NCOs in our Army painted as rapists of innocent, young, white women.

Mr. Warren said he wanted to leave no doubt that the NAACP wanted the guilty prosecuted. Dr. George said she wanted us to only charge the guilty. In another case where I should have kept quiet, I said, referring to the presumption of innocence principle, "We only charge innocent people, and then we let the judicial system determine if they are guilty or not." Her body language told me my comment was not well received.

We all agreed to work together to make sure everyone was treated fairly and the correct image of black NCOs was getting out to the public. In keeping with that, LTC Riesco was designated to work with Mrs. Grant on a joint USAOC&S/NAACP press release. We also agreed to meet again to review the situation. I told the delegation we would investigate the charges of the MPs' mistreatment of SSG Gamble. (After a lengthy investigation, including field grade officers sniffing car seats, it was determined by an AR 15-6 investigating officer from outside my command that SSG Gamble's allegations were baseless.)

After the NAACP delegation left, we held an after-action review. COL France was visibly hurt by the verbal attack on him. CSM Miller, who was black, said he was shocked at how the delegation's perceptions were not even close to the reality of the situation.

The meeting brought home to me that a lot of the concern about race was rooted in long-standing concerns by the African-American community. We had to be sensitive to that and make doubly sure no bias, either actual or perceived, entered into what we were doing. The issues were very emotional, and facts and numbers alone would not convince anyone we are being fair.

I knew back in October this day would come. It was much more difficult than I had ever imagined it would be.

9

★ ★

MULTIPLE AGENDAS TAKE SHAPE

WE PUBLISHED A PRESS RELEASE ON DECEMBER 20 detailing all the charges and specifications against Staff Sergeant (SSG) Delmar G. Simpson involving 25 female trainees and dating back to January 1995, when he arrived in the Aberdeen area.

An interesting exhibit at SSG Simpson's subsequent court-martial showed that when I had arrived at Aberdeen Proving Ground (APG), SSG Simpson's alleged sexual misconduct activity essentially stopped and then started up again in the spring of 1996. We caught him within 60 days. Almost 80 percent of his misconduct occurred before mid-August 1995, when I assumed command.

I recall reading a statement from another drill sergeant (DS) that quoted SSG Simpson as telling him, "I hate all women" and "I'm going to keep doing this shit until I get caught." He did.

On April 7, 1998, I was getting out of a van at Fort Hood, Texas, to have dinner at the home of MG Morrie Boyd, who was then the III Corps deputy commander. The civilian driver, Keith Conyers, asked if we could talk about the APG sex scandal before I got out of the van. I said sure. Keith said he had been a soldier at one time and had been in the same maintenance unit SSG Simpson was in at Fort Hood. He said SSG Simpson had been doing "the same stuff" in that unit. One day, SSG Simpson came in all smiles, and Keith asked him why he was so happy. SSG Simpson told him, "I'm going off to be a drill sergeant and have me a whole lot of fun."

Keith went on to say, "When I saw on TV about a sex scandal at Aberdeen, I turned to my wife and said, 'I bet Delmar is right in the middle of that.'" Much of what was happening to us in 1996–97 didn't make sense at the time, but as in the case of this example with SSG Simpson, a lot would become very clear to me before I retired from active duty on May 31, 2000.

At the Monday, December 23, update, it was noted that there was still an issue about medical and mental care for Reserve Component victims no longer on active duty. I was becoming frustrated. If Sara Lister, who was in charge of Manpower and Reserve Affairs, was so concerned with women's issues, it appeared to me the "Army's chief people officer" should be able to help us break through the log jam and get the issue resolved.

Ellie and I had a quiet but very nice Christmas. Remington was happy with all of his new toys!

Lieutenant Colonel (LTC) Gabe Riesco sent me an email at the end of day on December 27, saying he was finally able to make contact with Janice Grant of the National Association for the Advancement of Colored People (NAACP) to work on the joint press release. She had been on vacation. Mrs. Grant asked if we could wait until Leroy Warren returned from his vacation the next week so she and he could get together and work up a draft for us to review. We never received a draft from the NAACP.

General (GEN) Dennis J. Reimer sent an email to me and MG Thomas N. Burnette, Jr., commanding general of the 10th Mountain Division and Fort Drum, New York. It read:

> I would like the two of you to put together a pitch on sexual harassment. Bob, I'd like you to share with the divi-

sion commanders and TRADOC commandants lessons learned from your experience at Aberdeen. Tom, I would like you to put together a pitch which would be directed at early warning signs of harassment, discrimination, and abuse; how to detect them and how to overcome them. Would like [you] to give this presentation at the DC [division commander]/ TC [TRADOC commandant] Conference this spring [at Fort Leavenworth, Kansas]. Obviously, you two need to coordinate it, but I think both of you can develop it separately using the expertise you have available. If you need any help from us, please let me know. I'd like to give a take-away package to the people so that all of them can have the benefit of referring to that later and also benefit from the work you have done.

We also received an email from the Office of the Deputy Chief of Staff for Personnel at the Department of the Army (DA) directing: "Effective immediately all Army installations will begin collecting and submitting data on all sexual misconduct offenses to DA." Installations were further directed to report twice monthly through their Major Army Command (MACOM). In the case of USAOC&S, the MACOM was US Army Training and Doctrine Command (TRADOC) and the installation's was US Army Materiel Command (AMC). The installation would later state it would not consolidate the data for all of APG and that we would continue to report through TRADOC.

At the morning update on January 3, 1997, I announced the crisis action team (CAT) meetings would henceforth be held only on Mondays, Wednesdays, and Fridays in anticipation that our activity would be phasing down.

At the end of the day, I forwarded to all our higher headquarters LTC Riesco's nine-page after-action report on our experience of operating the DA hotline for 26 days. The report provided our assessment of what we did right and what we would do differently in the following general areas: facilities and equipment, media control, staffing, hotline operations, and miscellaneous. The report concluded with 23 observations and 11 recommendations for the DA to consider if it ever set up another hotline.

Saturday, January 4, was one of the worst days of my command tour at US Army Ordnance Center and School. At 1215, Colonel (COL) Dennis M. Webb called and said he had just been informed that a soldier was found dead of an apparent suicide in the barracks at Edgewood Arsenal (EA). COL Webb called back and said the deceased was the subject in an alleged trainee-on-trainee rape case dating back to July 20, 1996, and was scheduled for a court-martial hearing on January 7, 1997, the next Tuesday.

We convened the CAT, even though this case was not related to the ongoing investigation of DS/instructor misconduct with trainees.

We worked constantly throughout the day to determine why this happened. We notified all levels of command above us and worked with the DA Casualty Notification Office to ensure the family was notified before any information appeared in the media.

Over the next few days, both a medical and a psychological autopsy were performed, and we eventually knew pretty much conclusively what happened. This young male soldier was under a lot of pressure from sources not related to the military. That stress was coupled with a physical condition. We immediately arranged for counseling support for the soldiers at EA, and the battalion (Bn) commander briefed everyone on what had happened. I was not the only one shaken by this tragedy.

I was really concerned about Command Sergeant Major (CSM) Gerry Merrihew, who was visibly shaken. I asked COL Webb to arrange for his CSM, William Miller, to go be with CSM Merrihew. Gerry Merrihew really cared about soldiers.

On Sunday, I sent out an email to higher headquarters and the Army staff stating that due to media interest (16 inquiries), we had coordinated a press release on the apparent suicide.

This pointed out an ever-increasing problem—timely resolution of cases. Due to the sheer volume of work in the APG legal office, cases were taking longer and longer to reach adjudication. This, in my mind, only added to the stress some young men and women were experiencing.

I departed that afternoon for the winter TC/division commanders' conference at Fort Benning, Georgia. When we got to Fort Benning, I called back to receive an update from the team. LTC Riesco gave me a rundown on his interview with Channel 2, his three conversations with Mrs. Grant, and her interview on Channel 2 in Baltimore. I put this all into an email I sent the next morning to GEN William W. Hartzog, with copies to GEN Johnnie E. Wilson and the normal distribution at all my higher headquarters, Criminal Investigation Command (CID), and the Army staff. It read:

> Sir, Just want to give you an update since the last email on 20 Dec 96, on where we stand with the NAACP. Yesterday, WMAR, Channel 2, a local Baltimore TV station, sent us a fax outlining the fact that the NAACP has received numerous calls alleging that several Caucasians have engaged in sexual misconduct yet have not been charged. The fax also included a list of ten names along with the allegations associated with these names. We immediately sent this to CID and SJA.
>
> LTC Riesco made contact several times with Janice Grant, the president of the Harford County Chapter of the NAACP, to get any information on this list of names and assure her that we are aggressively investigating these leads as we do all others. Mrs. Grant indicated that she was not aware of who sent the information to Channel 2 and was satisfied we were taking appropriate action on the allegations.
>
> Channel 2 reported yesterday evening that the NAACP had called this information in and told [the station] that the Ordnance School was given a copy. They also pointed out that we took immediate action to pass this information to

CID. Through all of this we continue to maintain the moral high ground and will continue to work with the NAACP on our two common interests: equal and fair treatment of all soldiers and an impartial investigation. We also continue to work with the media to present a positive and proactive characterization of our actions. We continue to stay "in our lane" referring any investigative questions to CID PAO.... Will continue to keep you posted. V/R [Very Respectfully], Shadley

While discussing the suicide with MG James J. Cravens on Sunday, I said we really needed to move out and get legal actions completed in an expedited manner. He informed me "no legal timelines had been violated thus far on cases."

I got up extra early on Wednesday and made notes for my one-on-one meeting later in the day with GEN Reimer. I had charts to show him from the latest vector report: what happened, CAT composition and meeting frequency, actions being taken on the current situation, systemic causes that were emerging, preventive/corrective actions to preclude recurrence, emerging lessons learned, and tips for other commanders.

I took the morning update by phone in my room before departing for the conference. I approved the 15th edition of the weekly vector report, and the staff at APG emailed it out. I again stressed I did not think the current solution for how Reserve Component soldiers no longer on active duty were to obtain medical care was very convenient. The solution was for the soldiers to visit the nearest military facility, but many did not have a military medical facility nearby. It appeared to me the Army should just tell these women to go to their family doctors and send the bills to the Army.

LTC Riesco called and said Congressman John P. Murtha would be visiting us on Friday, January 10. This would be a big deal because he was the ranking minority member of the House National Security Appropriations Committee.

My meeting with GEN Reimer seemed to go all right. He agreed with me on the factors contributing to the situation in the Army. He said it was obvious that TRADOC had been cut too much, that the DS selection process needed to be improved, and that we needed to emphasize ethics to affect a change in the culture. He concluded the meeting by saying we were doing a good job and to keep working the issues.

While waiting at Lawson Army Airfield for the flight back to APG, I sat with MG Ken Guest and MG Bill Garrison, who was the commander in Mogadishu when the battle that subsequently served as the basis for the movie *Black Hawk Down* took place. I listened as they discussed the fight in Somalia. It made my problems seem miniscule.

The next morning, LTC Riesco gave me a message we had received from MG Morrie Boyd on Congressman Murtha's visit:

> Purpose is to gain 1st hand view of the sexual harassment situation from commander and drill sergeant perspective. Murtha has already visited [Camp] Pendleton (San Diego) and at later date may attempt Fort Jackson and Paris Island. He wants to verify his perceptions that we do not have large scale systemic issue that requires radical fixes to training base and to determine appropriateness of gender integrated training . . . See him as counter to Representative Livingston [R-LA] who feels we ought to roll the clock back on women in the military. Secretary of the Army is very supportive of this visit and he has requested to meet with Representative Murtha sometime after his visit to compare notes and to update [him] on current status of ongoing assessments. Representative Murtha will receive the legal brief regarding ongoing cases, etc. prior to visit. TRADOC alerted and comfortable with the visit.

COL Jerry Luttrell reported the DA inspector general inspection of our operations at Redstone Arsenal was going well. So far, it remained that only 2 of our 11 training locations—APG and EA—appeared to have sexual misconduct problems of significance. This was indeed good news.

Representative Murtha arrived at 0830. I met him and his party and took them to our command conference room in my headquarters. Accompanying the congressman were Greg Dahlberg, professional staff member on the House Appropriations Committee; John Plashal, professional staff member for the National Security Subcommittee, House Appropriations Committee; Mr. Horder from the Office of the Secretary of Defense; and LTC Cunningham from Office of the Chief of Legislative Liaison (OCLL).

The schedule called for an overview briefing and discussion from 0835 to 1000, but Murtha didn't want a briefing. He just wanted to talk. I sent the following report to all concerned at TRADOC, Combined Arms Support Command, CID, and DA:

1. Wanted to give you some feedback on our meeting with Representative Jack Murtha this morning. He arrived at 0830 with Mr. Dahlberg, Mr. Plashal, and Mr. Horner from OSD.

2. Overall it appears to have been a great visit. He didn't really want to see charts we had for him but we had an 1½ hour informal discussion with the brigade commander, two battalion commanders and their CSMs plus my staff and legal. Most of the discussion centered around dwindling resources and the quality of today's soldier. Discussion was open, frank, and candid, especially in the area of medical services available to soldiers, retirees, and families. (He was really surprised to learn we did not have a hospital at APG.) He seemed concerned with the resource cuts the training bases have taken and the impact it has had. Ended the session in a very positive note telling us what a great job we were doing for America's Army and said he wanted to tell the

drill sergeants the same. Mr. Murtha and his staff met privately with eight drill sergeants for about an hour.

3. Here's some of his comments to me at the end of his visit:

    - Sees the same things here as in the Marine Corps.

    - He appreciates the candor and openness of the leadership and the drill sergeants.

    - He expressed to the drill sergeants how important they are and the great job they are doing.

    - He's concerned about the impact of resource cuts.

    - Concerned about the short length of time to teach AIT (Advanced Individual Training)—too much to do, too little time.

    - Wants to make sure that we do not start to develop a "hollow Army."

4. Will yield to OCLL for an official report. Please let me know if there's anything else you'd like me to do.

On January 14, I sent the last weekly vector report. I announced in the cover email that the report would now be sent biweekly on the same dates as the biweekly TRADOC Sexual Misconduct Report. The objective was to ensure a single set of numbers on subjects and victims was distributed to avoid confusion.

We also learned GEN Reimer would be visiting APG in early February to meet with DSs. GEN Reimer and Sergeant Major of the Army (SMA) Gene C. McKinney had met with DSs at Forts Leonard Wood, Knox, Jackson, Lee, Eustis, Benning, and McClellan on January 6–8 "to demonstrate the Army Senior Leadership's pride in their professionalism and value to the Army and the Nation."

In an email to the Army leadership, GEN Reimer said, "As good as last year was for the Army, we are not perfect. Our problems with sexual harassment, initially identified in the training base, is a force-wide issue."

I again shook my head. Yes, sexual harassment is a wide problem, but the major problem was widespread sexual crimes—felonies.

In later correspondence, GEN Reimer would point out that over 2,000 DSs work hard every day to turn civilians into soldiers. When I was in school, if you scored 93 percent or higher, you received an A. If the Army got an A in DS performance, it would mean 1,860 DSs were doing world class work, but it would also mean 140 DSs were not doing what they should be doing. That's a lot of potential for trouble—about 10 problem DSs per gender-integrated training installation. This number was consistent with what we had discovered at USAOC&S.

There was an article by Rowan Scarborough in the January 14 edition of the *Washington Times* that discussed how the Marines conducted separate training for males and females and that "Some members of Congress, notably Appropriations Committee Chairman Rep. Robert L. Livingston, a Louisiana Republican, want the services to return to segregating male and female recruits, putting them under the control of same-sex drill instructors." This confirmed MG Boyd's comment about Murtha being the counter to Livingston's position.

At the Dr. Martin Luther King, Jr. Commemorative Prayer Breakfast on January 15, I sat with Janice Grant and again stressed my support for our mutual objective to be sure everyone was treated equally. I solicited her support to work with us to ensure the media portrayed what was happening as accurately as possible.

The guest speaker was Reverend Tatuem, who commented on the sex scandal as "not a skin issue, but a sin issue." To me, this hit the nail on the head. LTC Riesco would use this quote in a subsequent session with the media.

Court-martial charges for consensual sex against another DS, Sergeant First Class (SFC) Theron Brown, were preferred on January 17. (This case was subsequently resolved by the DS electing a discharge under Chapter 10 in lieu of court-martial, and the discharge was approved on January 24.) His rank of SFC (pay grade E-7) was reduced to Private E-1 (the lowest pay grade in the Army), and he was given an "other than honorable discharge."

We did have one female trainee who allegedly lied in her statement against a supposed subject, and we determined she should be court-martialed. This created a lot of media coverage, but it did demonstrate we were fair and considered all aspects of all allegations. Out of the hundreds of interviewed trainees, she was one of only two of who were thought to be less than honest.

On Saturday night, January 18, CSM Merrihew and I checked on soldiers in the barracks in the 16th Ordnance (Ord) Bn at APG. We found senior leaders present, working to make things better, and keeping order and discipline. The next night, we spent two and a half hours down at EA, going through the 143rd Ord Bn barracks. In discussions with DSs, we learned the processing of legal actions was still too slow from their perspective.

An interview with Mrs. Lister appeared in the January 20 edition of the *Army Times*. It was titled, "Lister: Future still bright for women in the Army." This was part of the DA's concerted media campaign to start spreading a positive word about the Army in general and opportunities for women in particular.

In an article in the January 21 edition of the *Wall Street Journal*, Adam G. Mersereau, Georgia State University School of Law, confirmed his position that separate standards for men and women in the military lowered morale and that Americans should be kept informed of the push for women in combat.

I would tell people that on any given day, someone could use the APG sex scandal to support any position regarding women in the military: no women in combat, women in combat, separate training for men and women, men and women trained together, no women in the military, the military would be better if women ran it, and many more agendas. This put us right in the middle because we were on point for the Army in regard to sexual misconduct. No matter how this all turned out, it would be our fault that someone didn't get their desired outcome.

We received an email advising that GEN Reimer had approved a sexual harassment chain-teaching package to be briefed to all soldiers by March 31. This is a technique where each boss teaches a class to subordinates, and then they teach their subordinates, until everyone in the organization has received the training. The program consisted of a video and a slide presentation to be given by the chain of command. This would shortly become problematic.

Since the beginning of the scandal, I had been presenting a briefing we developed on our situation and recommendations to preclude such situations to pre-command courses visiting APG for orientation. LTCs and COLs attended this course before taking command of a battalion or a brigade. This became a matter of routine for us, and I believe we gave these future commanders some very useful information. We received much positive feedback on this presentation from the participants.

Race, the role of women in the military, the image of the Army, and accountability were coming to the forefront as the agendas that would play out over the next several months and years.

10

★ ★

AGENDAS MOVE TO CENTER STAGE

EARLY IN THE AFTERNOON ON SATURDAY, JANUARY 25, Captain Paul Goodwin called to let me know we had received a letter via fax from Leroy Warren of the National Association for the Advancement of Colored People (NAACP). Mr. Warren's fax read:

> I want to thank you and your staff for taking the time to meet with the NAACP delegation [on December 19, 1996]. We deemed the meeting as being both helpful and informative. We left the meeting with the feeling of your personal concern for the overall issues discussed. We also felt that the Pentagon Counsel was basically out of the loop on the wrong side of the issues, involving race and racism. [Note: This refers to the comments made by the installation SJA Colonel Buzz France about the need to be sure we have a safe environment for "daughters" in the Army.]
>
> It was with great interest and absolute amazement that I was informed about midnight, on Friday, January 24, 1997, that there were some stunning events occurring relative to

the allegations involving the case against one or more black drill sergeants at APG. The 7:00 CBS Radio newscast (today) confirmed many of the rumors & allegations on the street. CBS Radio news named and discussed a [Pvt Doe]. [Note: I am not including the individual's true name as included in the letter. The other 7 victims, whose names appeared prominently in the media and/or court proceedings, are identified as Pvt Doe 2 through Pvt Doe 8 later in this book.]

We are somewhat disturbed and stunned because it seems that the Army in an attempt to limit negative publicity and damage control has possibly initiated a cover-up regarding [Pvt. Doe's] discharge. We have some concerns that APG is not being forthright and truthful with the media and the public. The word on the street is that a young white female from APG was recently treated at Walter Reed Army hospital for a possible life-threatening act. The failure to be truthful leaves many questions open to unhealthy speculation. The urgent and critical questions in our mind needing timely answers are as follows:

> Will the rumored forced release from the Army of [Pvt. Doe] impact the pending charges against any of the black drill sergeants?

> Have any white drill sergeants or white officers been charged during the current investigation at APG? Are any whites under investigation?

As I and the other NAACP officials told you and your staff during our meeting, we are not interested in protecting criminals and/or criminal activities. NAACP as an organization and we as individuals are interested in "JUSTICE, EQUALITY, and TRUTH." We want to work with the Army to ensure that JUSTICE prevails and only those guilty are charged. We are deeply concerned that innocent individuals, families, and careers are being destroyed or damaged for questionable reasons. Are all or some of the allegations attributed to [Pvt. Doe] and other complaints false? Are these allegations causing unjustified suffering to one or more black instructors? Have any blacks been questioned as to whether they have had any contact with the NAACP? NAACP will not accept a Third World anti-American witch hunt or personal terrorism of this type.

We urgently need to have a status update from the APG's information staff today or on Sunday (26 Jan 97).

Monday started out with several articles in the *Early Bird*. George C. Wilson wrote an article for the February 3 edition of the *Army Times* titled, "Cohen zeroes in on sexual harassment." New Secretary of Defense (SECDEF) William S. Cohen came to this cabinet position from the Congress as a senator from Maine. It was reported he said at his first Pentagon news conference he had "declared war on sexual harassment, warning that he would hold commanders at every level 'fully accountable' for any future abuses." He said, "There is a zero–tolerance policy as far as I'm concerned."

At the morning update, MAJ Susan S. Gibson reported the database on chapter discharges had been completed. Now all we had to do was cross-level that information with ongoing investigations to determine if there were any relationships. If a soldier had been discharged as a result of sexual misconduct by someone in the chain of command, then we needed to look at those cases. Charles Vickers, the 61st Ordnance (Ord) Brigade's civilian security officer, reported that Fort Sam Houston, Texas, had four courts-martial working for sexual misconduct. We later learned of the severity of these cases.

Senator Paul Sarbanes called from the Edgewood area where he was visiting and said he would like to come by my office. He arrived that afternoon with John Porter from his staff. I gave the senator a copy of the same briefing we had prepared for Congressman Murtha. It was essentially the same as we had presented to Congressman Buyer, Congresswoman Harman, and their delegation. We then talked about the situation at APG.

John Porter subsequently reported it had been a good visit and that the senator was amazed I was able to do all I was doing.

With the sex scandal demanding much of my time, I still had my primary job to do. I called General (GEN) William W. Hartzog, and we talked at length about my concerns with the new M88A2 Hercules heavy-recovery vehicle being developed to recover and tow the M-1 Abrams tank. This would continue to take almost as much of my time as the sex scandal. The civilians, warrant officers, and noncommissioned officers (NCOs) in our recovery training department had convinced me the Hercules could not safely tow the Abrams main battle tank (weighing 60-plus tons) up and down slopes on unimproved roads and/or cross-country under certain conditions. I concurred after watching demonstrations on a test course. This was not a popular position to take and would cause a significant challenge to the acquisition community. But I was convinced my concerns had to be addressed before I signed off on the safety release as the chief of Ord. The Hercules was eventually fielded and continues to perform well.

We published a news release informing the media that another drill sergeant, Sergeant First Class (SFC) William Jones, had charges referred to a special court-martial. The maximum punishment for this level of court-martial was six months of confinement, forfeiture of two-thirds of pay and allowances for six months, reduction to the grade of private (pay grade E-1), and a bad conduct discharge. The charges against SFC Jones included "seven specifications of failing to obey a lawful general regulation governing interaction between OC&S students and permanent party, one specification of being drunk on duty, and one specification of indecent assault."

The press release announcing SFC Jones's court-martial also reported on the discharge of SFC Brown in lieu of court-martial for the following charges: "one specification of sodomy, two specifications of failing to obey a lawful general order, two specifications of adultery involving one female soldier and one female civilian, two specifications of making a false statement under oath, and one specification of disobeying an order of a superior officer."

COL Dennis M. Webb and his team wrapped up a two-day cadre training course they had developed to help increase the professionalism of the school's cadre. This was just one element of a many-faceted get-well plan to correct things we found wrong during our investigations. I was pleased to see steady progress.

The Office of the Department of the Army Inspector General (DAIG) team inspected the explosive ordnance disposal training detachment at Eglin Air Force Base, Florida, on January 28. They had no negative findings to report. This was more good news and continued confirmation the sexual misconduct problems we uncovered at APG/Edgewood Arsenal did not exist at the other 9 Ord training locations.

I sent Major General (MG) James J. Cravens an email on January 29 thanking him for what Training and Doctrine Command (TRADOC) had done so far regarding holdovers. I invited him to have the TRADOC surgeon talk to our commanders and look at the health care system and our holdover situation. It appeared to me the medical folks were trying to operate the system like a business focused on saving money rather than helping customers. But efficient medical support did not equate to effective medical support for commanders with soldiers waiting at APG for weeks for medical appointments in Washington, DC.

In a January 30 article in the *New York Times* by Eric Schmitt titled "Army Inquiry Found Early Evidence of Sex Harassment," it was reported that in a 1996 study of extremist activity, numerous complaints of sexual harassment were uncovered. The article stated the complaints were referred to local commanders and were not included in the final report under orders from Secretary Togo G. West. Schmitt wrote, "Indeed Mr. West has acknowledged since the Aberdeen scandal broke that the Army's top leaders may have been lulled into thinking they had the problem under control, inasmuch as Congress and independent experts had praised the service for its sexual harassment policies." I had not heard before that Secretary West had directed modification of the DA inspector general's final report. This would apparently not be the only time he directed the modification of a DAIG report.

At the morning update on January 31, MAJ Gibson announced that charges on Sergeant (SGT) Isiah Chestnut, an instructor and not a drill sergeant (DS), would be referred today to a special court-martial. The charges against SGT Chestnut consisted of four specifications of failing to obey a lawful general regulation governing interaction between US Army Ordnance Center and School (USAOC&S) students and permanent party and one specification of assault. These charges were the result of a call to the hotline and resulted in our sixth cadre member being formally announced for alleged misconduct.

Ellie and I closed out the first month of 1997 on a fun note. We drove down to Fort Myer, Virginia, to spend the night with Lieutenant General (LTG) Thomas G. Rhame and his wife, Lin. We attended the retirement ceremony and reception for LTG Samuel E. Ebbesen and his wife, Lillian, whom we had known since we had served together at Fort Campbell in the mid-1980s and lived near each other in Springfield, Virginia, when Sam and I both served in the Pentagon. I appreciated both LTG Rhame and LTG Ebbesen looking out for me.

The *Washington Post* had an article in its February 1 edition in response to our press release about SGT Chestnut. The article also included, "At Fort Leonard Wood training center in Missouri, officials said this week that 28 men were under investigation for allegations of misconduct. Two others have pleaded guilty and four await courts-martial." I thought, "Wow—and *we* are the center of attention."

Later in the day, we received a fax from TRADOC saying GEN Reimer's chain-teaching program had been put on hold by the Army's deputy chief of staff for personnel. In bold type in the fax said: "Do not copy video and do not disseminate the video." This was strange, but it would soon become clear why this action was being taken.

I spent the afternoon with COLs Tom Leavitt and Thomas Lopresti from DAIG giving sworn testimony in their investigation of who knew what and when. I made a huge mistake by not taking MAJ Gibson with me. I would never again give a statement to any investigator without a lawyer present. I was honest and straightforward but probably too short in a few of my answers.

For example, I remember COL Leavitt pointing out the window to a building across the street and asking if I had gone through it when I did my initial assessment of the command when I assumed command. I simply said no. I should have said, "No, because that building was under renovation at the time and the interior was gutted."

I also made the comment that I thought the medical community was trying to run their operation like a business saving money. I later learned this may have been taken that I was running the command like a business. Either that, or someone else they interviewed may have said the commanding general ran the school like a business, but was referring to the previous commander. According to people at APG at the time, he did just that.

The problem general officers have when they are being investigated is they are not authorized to know who else was interviewed by the DAIG and they are not allowed to see anything said about them. COL Leavitt did tell me (as contained in the official transcript), "You are not suspected of any criminal offense and are not the subject of any unfavorable information." The first part turned out to be true, but not the second. My interview supposedly ended the DAIG's investigation of the chain of command at APG. This turned out not to be the case, however.

I was not very pleasant to be around after my testimony, and Ellie let me know it as we celebrated our wedding anniversary that night at dinner. I was out of the running for husband of the year right out of the blocks.

MAJ Gibson told me later generals make lousy clients. They know they are right and they talk too much. She had me pegged. I kicked myself again for not taking her with me to the interview.

The front page of the *New York Times* on February 4 contained an article by Eric Schmitt: "Top Enlisted Man in the Army Stands Accused of Sex Assault." Retired Sergeant Major (SGM) Brenda L. Hoster had come forward publicly and accused Sergeant Major of the Army (SMA) Gene C. McKinney of multiple incidents of sexual misconduct, including sexual assault. (Note: I am using the victim's name here because she was not under my command and I did not control her information. Also, as noted in the media, SGM Hoster voluntarily made her name public.) She also claimed she had reported the incidents to her chain of command and nothing had been done about them. The story was also in the *Washington Post* and carried on CNN. From the photos in the media, everyone could see SGM Hoster was white and SMA McKinney was black. This would subsequently feed the racial implications of the Criminal Investigation Command investigations.

I now suspected SGM Hoster was the anonymous caller to the hotline making a complaint about SMA McKinney the day I had been in the media center. We now knew why the chain-teaching program was put on hold and the video was embargoed: SMA McKinney was in the video.

We put out a press release on the preferring of charges on another DS, Staff Sergeant Vernell Robinson, Jr. The press release also announced, "There have been seven OC&S personnel charged since November when the Army made the allegations public. In addition three others have received non-judicial punishment in the form of fines or, in one case, reduction in grade for non-criminal offenses such as inappropriate comments. A fourth non-judicial action resulted in a finding of not guilty in an allegation of sexual misconduct with a trainee."

I was pleased the Test and Evaluation Command public affairs office put in the last paragraph of the release: "The Ordnance School continues to follow the three primary objectives set by its commander, Major General Robert D. Shadley, at the beginning of the investigation: identify and care for the victims; investigate all allegations to the fullest extent and pursue due process when required; and identify systemic causes and apply corrective action to prevent further instances of misconduct."

The old saying "It's hard to remember that the mission is to drain the swamp when you're up to your butt in alligators" was a constant reminder to keep everyone's focus on the mission, and this paragraph helped.

John Porter provided me a short summary of the testimony by Secretary West, GEN Reimer, LTG Jerry Bates, and Brigadier General Daniel A. Doherty that had occurred earlier in the day before the Senate Armed Services Committee (SASC). Senator Snowe picked up on SGM Hoster's allegation that her chain of command did nothing, using it as another example of how it appears the Army covers up such misconduct. She went on to say that initial reports at APG were not acted upon. John also reported Senator Snowe did not think women were safe in the Army. Senator Levin questioned whether SMA McKinney should be on the Army panel investigating sexual harassment. Secretary West replied he had just found out about the SMA McKinney allegations. Senator Robb raised the issue of whether or not women and men should be trained together.

John went on to talk about the need for numbers of victims and subjects because Snowe told a reporter she understood 50 percent of the DSs at APG were suspended. Apparently the Senate staffers were not well informed on several issues and had passed bad information on to the senator.

Paul Boyce sent me the written statements made by GEN Reimer and Secretary West. GEN Reimer's comments were essentially the same as he had made in the past. Secretary West's statement really bothered me. He started by saying, "I welcome the opportunity to appear before you today to report on the recent incidents of sexual misconduct at Aberdeen Proving Ground." He went on to say, "What is alleged to have occurred at Aberdeen was particularly troublesome to us because it involved abuse of authority and it appeared that the incidents either had gone unreported or were not addressed."

I concluded his comments about the chain of command fed Senator Snowe's perception of what had happened. Secretary West did talk about the DAIG investigation and his panel, but the tone was that they were ensuring that what was going on at APG was not going on anywhere else.

The Army leadership's overall strategy of dealing with an Army-wide sexual misconduct problem was emerging: Only APG was a problem.

After Senator Snowe mentioned an incorrect percentage of suspended DSs, MAJ Gibson and I worked the actual numbers: 10 suspended DSs equaled 12 percent of the total DS population; 10 suspended instructors equaled 3 percent of the total instructor population; and the total cadre suspended equaled 1.5 percent of the total cadre population. This numbers issue would go on for weeks and would take on a life of its own.

GEN Reimer, in response to a question at the hearing, stated he thought it might be time to reexamine gender-integrated training. His comments were a page-one story in the *Washington Post*.

In an article by Eric Schmitt in the *New York Times*, Senator Snowe's comments about women not being safe in the Army were reiterated along with her comment, "It's not just sexual harassment. It is abuse of power and sexual misconduct." The senator was right on.

In the following days, Army personnel would make many comments in the media to the effect of, "What GEN Reimer meant to say was, women have a very valuable place in our Army, and we are going to continue gender-integrated training."

I talked with MG Morrie Boyd about the SASC hearing, and he confirmed that a lot of the discussion focused on gender-integrated training and the numbers of victims and subjects.

Morrie went on to say the Army was content to have APG be the focal point. This softened a little bit of my feelings about Secretary West's statement, as I understood Morrie's comment to mean we were doing a good job of showing the world we were doing the right things in a tough situation. It did, however, still look as if we were about to, as LTC Riesco said, take one for the team.

I called GEN Hartzog at the end of the day and told him what information I would be giving GEN Reimer when he visited on February 8. I also told him I did not think the DAIG would give me the information I needed on the accountability of the chain of command so I could take action.

At the morning update on February 7, I learned we could expect a major story coming out of Germany about sexual misconduct in Darmstadt. It seemed every 7 to 10 days, another sizzling military sex scandal story would hit the media. I was beginning to believe this would never end.

The Office of the Judge Advocate General at the DA requested from COL France the weekly by-name data on all sexual misconduct allegations at USAOC&S so it could send details to Congress. This was bizarre. Now we had staffers and members of Congress looking at Article 15s.

I sent MG Cravens an email letting him know about this apparent violation of the chain of command. I also told Jim it appeared no one was reading the TRADOC report I sent him biweekly that was now up to 16 pages and that gave the same information. I was now faced with the requirement to submit a weekly and a biweekly report going to different places with essentially the same information. I concluded by saying, "Would anticipate Congress will break the code that this is bigger than just USAOC&S and others will be asked for similar data."

COL France sent a long email on the issue of providing data to Congress, and I sent it along to MG Cravens. Buzz hit the nail on the head: "The problem is that [the House] does not want to be scooped by the media or the Senate. Congressional interest has obviously shifted from understanding the systemic causes of the problem to very close scrutiny of how we are dealing with individual cases."

GEN Reimer cancelled his February 8 visit because of poor weather. He called me at home and apologized for not coming. We talked about the chain-teaching program, and I said I would send him an email with my observations. He went on to say there was a fine line between warfighting and respect, but that we were doing the right thing.

Paul Boyce sent me a long fax on Sunday that contained the upcoming front-page story in the February 17 edition of *Army Times* about SMA McKinney. There were three other *Army Times* articles in the fax, and the last one caught my eye. An article titled "New doubts circle Army's ability to probe harassment" by C. E. Willis brought up the issue of the of effect SMA McKinney's dismissal from Secretary West's panel. The last paragraph stated, "Judith Youngman, professor of political science at the U. S. Coast Guard Academy, is listed as a consultant to the Army panel in her capacity as chair of the Defense Advisory Committee on Women in the Services. However, she said, 'Although I am named on the panel as a consultant, I've never been contacted. I guess that kind of speaks for itself.'" Comments such as this added to my concern about how objective the panel would be.

In an article in the *New York Times*, Susan Barnes was quoted as saying, "At some posts, they are not reporting the rapes. Post commanders don't want those statistics." I have gotten to know Susan over the years, and she was absolutely correct in her statement. I had one fellow two-star commander tell me I was stupid for reporting rapes. I would later be told TRADOC headquarters apparently knew he was not reporting all the sexual misconduct on his installation.

As predicted, there was a long article in the European *Stars and Stripes* about three NCOs being suspected of multiple offenses, including the rape of students at a two-week course at the Darmstadt In-Processing Training Center. This story would continue to develop and was known in Europe as "Little APG."

11

★ ★

OUR SOLDIERS ARE
TAKEN ADVANTAGE OF

I FLEW TO REDSTONE ARSENAL, ALABAMA, ON THE AFTERNOON of February 10, 1997, to conduct a review and analysis with Colonel (COL) Jerry Luttrell and his team. I would also present the chain-teaching program to them on February 11.

I was notified that General (GEN) Dennis J. Reimer would be visiting us on February 12. We would follow our original plans for his visit, so even with this short notice, it would be no problem to accommodate his schedule.

I met GEN Reimer at the helipad on February 12 and escorted him to Dickson Hall, our auditorium. He spoke to over 250 leaders, including our drill sergeants (DSs), and he fielded questions from the audience. I then had a 30-minute office call with GEN Reimer before his flight back to the Pentagon. I reviewed the scandal-related actions we were taking and then covered all the other things we were doing simultaneously to improve technical training. He told me he was glad to see I was working on important things other than the sex scandal.

I sent GEN Reimer a note to thank him for his visit and let him know I had gotten a lot of positive comments about his presentation. He responded, "Thanks Bob . . . quite frankly an uplifting experience for me . . . appreciate the job you are doing keeping the focus on what is really important and enjoyed meeting some of your key people."

I called Major General (MG) Ken Guest that night to give him a personal report on the chief of staff of the Army's visit. MG Guest said he had talked to GEN William W. Hartzog and that the Department of the Army (DA) inspector general team would brief him the next day on the who-knew-what-when question.

MG John M. Longhouser called with some good news. He had been notified by Chaplain (MG) Donald W. Shea, the Army's chief of chaplains, that two chaplain teams had been approved for the 61st Ordnance (Ord) Brigade. This now meant both the 16th and 143rd Ord Battalions (Bns) would have full-time chaplain support available for the trainees as another avenue for trainees to let us know about problems.

On Tuesday, February 18, MG Longhouser, as the general court-martial convening authority, approved a Field Grade Article 15 instead of a court-martial for Staff Sergeant (SSG) Beach. Even though I was not in the legal chain, I understood the legal community's concern about the viability of Pvt Doe as a witness.

The Aberdeen Proving Ground (APG) staff judge advocate general sent weekly updates on investigations and court dates. The update published on February 19 included: Sergeant Chestnut requested and received a discharge under Chapter 10 in lieu of a court-martial; SSG Beach's case would now be resolved by Article 15 proceedings; and Article 15 proceedings held for three instructors involving non-felony offenses resulted in two not-guilty verdicts and one withdrawal of action because the instructor had received prior punishment for the offense.

As anticipated, the announcement did get media interest in a *Washington Post* article by Jackie Spinner. The focus was on SSG Beach now going before an Article 15 proceeding. Karen Johnson, vice president of the National Organization for Women, was quoted as saying the decision "shows again there really is a lack of will to take this seriously. They are not willing to really enforce infractions of zero-tolerance policy." To me, the referral of SSG Beach to an Article 15 proceeding meant the judicial system looked at each case on its merits and took action on each one as appropriate.

Major (MAJ) Susan S. Gibson reported at the morning update that another female trainee, Pvt Doe 2, had charges against her referred to a summary court-martial for multiple alleged offenses including false statements in a sexual misconduct case.

I called GEN Hartzog on Saturday and talked with him about a myriad of subjects, including the Beach case. GEN Hartzog said GEN Reimer had ordered the judge advocate general of the Army (MG Mike Nardotti) to have a judge or lawyer look at the Beach case. I continued to be amazed at how the Army leadership—both military and civilian—would bend in whichever way the winds from the media or Congress blew. Here we had the highest ranking officer in the US Army worrying about punishment of a sergeant six levels of command below him.

The weekend newspapers contained two scandal-related articles, which again made the point that sex sells.

Tom Bowman's article, "Accusers Consented," in the February 23 edition of the *Baltimore Sun* discussed "an extensive breakdown in discipline and a freewheeling, college-like environment at the U.S. Ordnance Center and School" where "40% of female recruits alleging abuse admit some contact willing." This article then continued with the debate of whether trainees could have consensual sex with their DSs and instructors. I had already repeatedly stated my position that there was no such thing as consensual sex in the case of DSs and trainees.

An *Associated Press* article in the February 22 edition of the *Washington Post* reported that the number of victims in the Darmstadt, Germany, scandal was now up to 21 for the 3 noncommissioned officers (NCOs) alleged to be the perpetrators. Again, this made the case that APG was not the only place where sexual predators had multiple victims.

I asked Lieutenant General (LTG) Jerry Bates for a copy of his report that assessed the responsibility and accountability of the chain of command. I needed this information to close out actions and move on. I was also sure from my reading of the statements that some NCOs and company grade officers knew what was going on and did not report it. I garnered from my readings that this was common knowledge in the lower ranks. If we did not do something to those who failed in leadership, we would be accused of letting people off the hook.

LTG Bates responded to my email request for the results of his team's investigation of who knew what, when. It read: "Bob, understand your request. My report will go to SECARMY [Togo G. West] first and then, based on his approval, to GEN Hartzog. So it will be a while before it is available for your use." I never received his report or anything about accountability.

COL John A. Smith called on February 26 and said Channel 9 from Washington, DC, was on the way to APG to talk to some sergeants who wanted to meet with the press. The NAACP held a press conference calling for an investigation because black DSs were being targeted and all the accusers were white. I watched Mrs. Grant and Dr. George on Channel 9 at 1700 hours.

The Thursday, February 27, edition of the *Baltimore Sun* contained Mrs. Grant and Dr. George's allegations that white women were reporting black men. This was, as Dr. George said in the article, "bringing back images of black men walking down the street and saying 'hi' to a white woman and then being hung for it." The article went on to say the local NAACP chapter was calling for an independent investigation.

I continued to be disappointed that the civilian and military leadership at the DA didn't give us more "top cover." The senior leadership in the Pentagon had all our data and more, as they had direct access to all the CID's detailed information. They knew we were not targeting blacks or anyone else, and they knew the victims represented the overall ethnic makeup of the Army. To my knowledge, the Army senior leadership provided absolutely no help to us with regard to the alleged race issue.

COL Tom Leavitt called and reported what I already knew—the leadership at Fort Leonard Wood did not want to interview former students as we had done. I told him approximately 40 percent of our victims had been identified by interviews with former students, and if the Army wanted to help victims and identify perpetrators, it should do such interviews.

I continued to have a hard time trying to figure out why the Army did not want to interview former students at Forts Leonard Wood and Jackson. I would later conclude, based on a good friend who had worked in the Pentagon during this time, that the Army leadership felt any more interviews would be a self-inflicted wound. The Army was still dealing with over 1,000 cases from the hotline at places other than APG, and it was very convenient to let us be the sole lightning rod for the Army catharsis. This friend had also attended a high level meeting with GEN Reimer where the judge advocate general (MG Mike Nardotti) told GEN Reimer the Army was in a "self-destruct" mode. This may have been why the Army didn't do further trainee interviews.

This to me was very disturbing. It appeared the Army leadership was more concerned about their own image and the Army's than about locating and caring for victims and finding perpetrators who should no longer be leading young people in our Army.

The media blitz continued. The Monday edition of the *Baltimore Sun* had a front-page story by Scott Wilson, "Aberdeen sex cases could put race on trial." It kept the race issue front and center. Janice Grant was quoted as saying, "I'm at the point now where I'm skeptical of what is being told to us because so much of what is being told is turning out to be 100% untrue." It would be months and years before the truth in the NAACP claims would come out, but getting to that point was very painful for many of us at the receiving end of such false allegations by Grant and the NAACP.

The NAACP and then the Black Caucus would continue to say they didn't understand why so many of the subjects were black. One of our company commanders, Captain (CPT) Alicia Jackson, who was black, was quoted in Scott's article: "The majority of the drill sergeants are ethnic and minority, so when these charges come up the probability of them being against someone black or ethnic is greater." CPT Jackson was commander of the 143rd Ord Bn's Alpha Company. "If they did the act, then now they have to pay the price," she said.

There was also an article by James Brooke in the *New York Times* on March 3 contrasting Susan Barnes and her group, Women Active in Our Nation's Defense (Wandas), and Elaine Donnelly, president of the Center for Military Readiness (a conservative policy group in Michigan). Brooke wrote: "Mrs. Donnelly contends that sexual harassment in the military will decline when the Pentagon declares that 'the social experiment is not working' and withdraws women from joint training and joint housing with men." Ms. Barnes's group provided legal assistance for women in the military pro bono through the Wandas Fund, and her organization also had a lobbying subgroup called Wandas Watch to work the Washington political scene. Wandas supported increased roles for women in the military. Susan Barnes was Sergeant Major Brenda L. Hoster's attorney in the case against SMA McKinney.

These two articles again focused attention on the two main agendas at work—race and women in the military—and put us right in the middle as the poster child for the Army.

On March 4, MAJ Gibson called to let me know Channel 2 wanted to come to APG to talk with a female trainee who wanted to recant her story. Pvt Doe 2 had allegedly told the media she was recanting the statement she was pressured into making. We put out a press release that read:

> NOTE: Normally information concerning charges in a Summary Court-Martial are not made public. However, due to media interest sparked by comments made by PVT Doe 2 on March 4, the following is provided to aid the media in their coverage.
>
> Charges we preferred on Feb. 20 against a private who made allegations as part of the ongoing investigation of alleged sexual misconduct here. Charges were referred to a Summary Court-Martial on Feb. 28.

PVT Doe 2, a former student at the U. S. Army Ordnance Center and School here, has been charged with being absent without leave, disobeying the order of a noncommissioned officer, making a false statement under oath, and breaking restriction.

Her case will be heard at an as yet unscheduled Summary Court-Martial, which is the lowest level of court-martial. The maximum punishment that may be adjudged is reduction to the lowest enlisted grade (Private), two months restriction, 45 days extra duty (if confinement is not adjudged), forfeiture of two-thirds of one month's pay, and confinement for 30 days.

At the morning update, Chief Warrant Officer 3 (CW3) Don Hayden reported that a retired sergeant first class had been added to the list of subjects. He retired in 1994, and we would later learn he was another one of the senior DSs who introduced Playing the Game to new DSs at APG. This Game connection would be solidified along with an understanding of the race issue when one of the DSs, who had 10 victims, decided to work a deal with MG Longhouser. He agreed to become a witness for the government in return for a lighter sentence. His statements would put everything into perspective by the end of May.

At the morning update on March 6, we talked about the status of the four victims still at APG. Pvt Doe 3 and Pvt Doe 4 wanted to move to their next assignments, Pvt Doe 5 wanted to stay at APG, and Pvt Doe 2 was awaiting court-martial. These were all legal decisions involving how to best handle witnesses. I recommended we send Pvts Doe 3, 4 and 5 on to their next units and then call them back for trials as needed.

On Monday, March 10, we published a press release about the eighth cadre member at US Army Ordnance Center and School being charged: "The Army preferred six charges against Staff Sergeant Herman Gunter. . . ." We went on to report that of the seven previously charged, four were awaiting courts-martial, two were discharged under the provisions of Chapter 10, and one was now having his case handled under

Article 15. We also updated the media on the six non-felony cases adjudicated via Article 15: Three were found guilty of non-felony offenses and punished; two were found not guilty; and one was found to have already been punished for the infraction.

The Channel 11 late news announced Janice Grant, president of the Harford County NAACP, would be holding a news conference at her home the next day, March 11, with four of the alleged victims who would recant their allegations against DSs and say they were coerced into their statements. They would also claim they were promised specific assignments if they made statements against DSs but that the promises were not kept.

The announcement of the NAACP news conference was carried on CNN at 0600 hours on March 11. An *Associated Press* article led with "Four female privates at the Aberdeen Proving Ground will recant their rape allegations against instructors at the military training facility, an NAACP official said yesterday."

I spent all morning calling people in the Pentagon, CODELs, CID, legal, public affairs offices, and GEN Hartzog about the NAACP news conference. My administrative assistant, Carol Nye, came into my office and said Representative Molinari was on the phone. I picked up the phone, and she said, "General Bob, I'm going on CNN today. Anything you would like me to say?" I told her this was an issue of sin, not skin, and I gave her ethnicity data. I also gave her some other data points and discussed the situation in general. She later would use some of what I gave her, but not all. I thought this was a class act on her part. Other than Senator Sarbanes, Senator Mikulski, and Representative Ehrlich, she was the only person working on Capitol Hill (member or staffer) who ever called to check facts with me personally.

The news conference started at 1300 and was carried live on CNN. Kweisi Mfume, president of the NAACP, initiated the session by calling for an independent investigation because, based on the information provided by the female trainees present, there might be a violation of civil rights. Janice Grant followed Mfume and introduced Stewart

Robinson, counsel for the Harford County Branch. Mfume came back to the microphone and praised the trainees' courage for coming forward. He also noted they had contacted the NAACP for assistance.

Pvts Doe 2, Doe 3, Doe 4, and Doe 5 were then invited to speak. Former Pvt Doe 6, who was now a civilian, also spoke. None of the five recanted a rape charge because none of the five ever alleged they were raped. Their complaint was that the CID investigators pressured them to allege rape. The four still on active duty were upset they had not been moved on to their next assignment.

In the question-and-answer session that followed the statements, Pvt Doe 6 said, "We never said we were raped. Nobody was charged with rape." Pvt Doe 3 said, "They wanted us to say that, yes, but they can't charge them [drill sergeants] unless we agree that it was rape, and none of us said that." She went on to say that other trainees had been raped, but not any of the five at the press conference.

A reporter asked, "Did you all sign statements saying that you were raped?" Pvt Doe 3 replied, "We never signed statements . . . we didn't . . . I don't know . . . none of us have signed statements saying that we've been raped."

In response to another question asking if they signed any papers, Pvt Doe 3 later said, "We signed statements. We gave statements. We were coerced into making statements. However, the statements that we made were the truth."

As I watched this 45-minute session, I thought, "This is craziness. It doesn't make any sense." It would all become clear a couple months later, when, as a part of my own investigation of the chain of command, I read every word of every statement by every alleged subject and victim.

One of the DSs thought it would be good idea to get Pvt Doe 2 to recant her statement. He gave her Mrs. Grant's phone number, and Pvt Doe 2 and Grant met. The DS then got Pvt Doe 3 to do the same thing. Two of his fellow DSs contacted Pvt Doe 5 and Pvt Doe 6 to participate and gave them Mrs. Grant's number. These were the original four who were announced as going to be at the conference. These four trainees took the scheme to two other trainees. Pvt Doe 4 agreed to participate, making her the fifth alleged victim at the news conference. The other trainee, Pvt Doe 7, who was being treated for emotional problems, declined to participate.

I understand Mr. Mfume was upset before the news conference because it was clear to him the trainees had made no rape charges to recant, so they were not eager to go on TV. I would soon learn the trainees had been promised free legal services if they made a statement for the NAACP. Jackie Spinner of the *Washington Post* and I talked about this before I left APG. She had heard most of what had gone down in that conversation among the NAACP officials on March 11, as she was in Mrs. Grant's bathroom prior to the press conference beginning.

This was a terrible situation. These five young women were exploited first by the DSs and then by the NAACP. This is a prime example of why young people desperately need strong, positive leadership. Teenagers make poor decisions on occasion, and they need adult leadership, not exploiters like sexual predator DSs and organizations with an agenda.

Poor Brigadier General Daniel A. Doherty spent weeks rebutting the charges that his investigators did anything wrong. He told me later one of his folks printed the photos of the investigators on a single page in order to, without a word, show that the agents doing the interviews were young and were a mix of gender and ethnicity.

At the morning update on March 12, COL Dennis M. Webb reported that the four Pvts at the previous day's press conference would leave APG for their next assignments. All the moves were as planned at our morning update on March 6 before the news conference and not in reaction to the news conference.

Pvt Doe 3 asked if she could see me, and I said sure. I had Command Sergeant Major Merrihew sit in with me. She said the NAACP had told her that if she came to Mrs. Grant's house and made a statement, she would be provided free legal help. She had gotten scared after talking with CID investigators, and it sounded like a good deal to get free legal help. She also said she did not like the CID investigative techniques and felt she had been taken advantage of. She said she was looking forward to going to Fort Riley.

Pvt Doe 3's final comment really struck home with me: she said the DSs did everything for them, and they were wonderful. When they asked for sex, she thought she owed it to them. It wasn't rape, and it wasn't consensual. She concluded with, "There's a big gray area in between."

I told her to go to Fort Riley and enjoy the assignment. I also told her to have faith that everything would work out okay. She said she was sorry if she had said anything wrong on TV. I did not know at the time a DS had talked her into going to the news conference. It would not have changed my mind that here was a young kid who had been taken advantage of by several adults. I told her she would not be punished for speaking her mind.

On March 12, Leroy Warren, Jr., chairman of the national NAACP Criminal Justice Committee, sent a letter to Secretary of Defense Cohen. The first paragraph contained a resolution the NAACP national board of directors passed on February 15: "The NAACP National Board of Directors directs the National Criminal Justice Committee and the NAACP National board to respond to what seems to be a willingness by the U. S. Department of Defense, especially the U. S. Army to punish blacks alleged to be guilty of SEXUAL HARASSMENT, but MINIMIZE similar or complaints against White Officers." (Accented words in original letter.)

The letter called for an external investigation because of concerns about civil rights violations and cover-up by CID. I found one sentence ironic: "The word on the street is that the U. S. ARMY is in the process of redeploying the white females who recanted/denied the charges that they were raped by black Drill Sergeants. Were the statements made by the women today at the press conference the basis for their rumored transfers?" The trainees never claimed rape!

COL Smith called on March 13 and told me there would be no independent investigation as sought by the NAACP. However, Steve

Buyer (R-IN), chairman of the House National Security Committee Subcommittee on Military Personnel, had published a press release. It read:

> In response to tremendous Congressional concern over the recent allegations of sexual misconduct in the U. S. Military, Speaker Newt Gingrich has tasked the House National Security Committee, the committee with principal jurisdiction over the Department of Defense, to lead Congress's efforts to ensure that such misconduct is fully investigated and all appropriate actions to prevent future abuse are taken. Chairman Spence has asked me, as chairman of the Military Personnel Subcommittee, and Tillie Fowler as one of the senior-most women on the committee, to lead the committee's efforts on this very important matter. Representative Ron Dellums, the committee's ranking member, has asked Jane Harman to assist us in our endeavor. In that capacity, we have begun a bipartisan, systematic, and thorough effort to understand the causes and contributing factors that led to the events at APG, as well as other training centers. It is an on ongoing effort that will look at each of the military services, not just the Army. The focus of our investigation will be the issues of sexual misconduct, sexual harassment, and fraternization.

The press release went on to announce this would be a hands-on investigation with members of Congress visiting installations, starting with Fort Leonard Wood the next day. I was glad to see Buyer recognized the problem was more than just sexual harassment and included more than just APG. But I did not hold hope that he and his group would be of much value in solving the problem of sexual misconduct in the Army. I would be proven right.

12

★ ★

CONGRESSIONAL VISITS CONTINUE

O N FRIDAY MORNING, MARCH 14, 1997, WE WERE NOTIFIED that the Congressional Black Caucus would visit us on Monday, March 17, and that Senator Snowe would visit us on Friday, March 21.

The *Washington Times* reported that Secretary of Defense William S. Cohen had met with General (GEN) Dennis J. Reimer and Secretary Togo G. West to discuss the Aberdeen Proving Ground (APG) scandal. Secretary West said the Office of the Department of the Army Inspector General (DAIG) would do the investigation of the race issue once the criminal investigation was done. Cohen was quoted as saying, "I have no reason to conclude that there were [racial motivations]."

A *Baltimore Sun* article by Tom Bowman reported that members of Congress had refused requests by the National Association for the Advancement of Colored People (NAACP) and the Congressional Black Caucus for an independent investigation outside of the Army. This confirmed what Colonel (COL) John A. Smith had told me.

An article by Dirk Johnson in the March 17 edition of the *New York Times* discussed gender-integrated training at the Great Lakes Naval Training Center. The thrust was that mixing men and women in equal numbers in training divisions was working well. The counter to this positive story about women in the military was an article in the *Washington Times* by Rowan Scarborough, which reported: "The Army is circulating a handbook to unit commanders that says deployed female soldiers are more prone to injuries and fatigue than men, risk dehydration because of reluctance to use public latrines and should eat two-thirds rations to avoid gaining weight." The pros and cons of women in the military and gender-integrated training continued to play out in the press.

The leaders at the Great Lakes Naval Training Center would learn eventually that "fame is fleeting" as they would also have sexual misconduct cases involving multiple cadre with multiple trainees.

The five members of the Congressional Delegation (CODEL) we hosted on the afternoon of March 17 were Rep. Maxine Waters (D-CA), with staffers Joseph Lee and Donna Crews; Rep. Elijah E. Cummings (D-MD), with staffer Anthony McCarthy; Del. Eleanor Holmes Norton (D-DC); Rep. Juanita Millender-McDonald (D-CA); and Del. Donna Christian-Green (D-VI).

Also in attendance were four members from the Department of the Army's Office of the Chief Legislative Liaison (OCLL): Lieutenant Colonel (LTC) Scott C. Black, LTC Tom Hawley, Dorothy Groome, and Staff Sergeant (SSG) Carol Murray.

Before the meeting began, it was noticed that Janice Grant from the NAACP had invited herself to the briefing for the CODEL. Representative Waters appeared not to be comfortable with Mrs. Grant in attendance and asked her very politely to leave. I would later wonder if Representative Waters knew then what Ed Starnes sent me in an email on July 22, 1999, after I had been gone from APG for over two years. The email quoted an article in the *Record*, a weekly newspaper published in Bel Air, Maryland, for the residents of Harford and Cecil Counties:

> Janice Grant has been charged with not filing financial records with National since she has become President of the Harford County chapter. She, according to a paper dropped

off at The Record, has not been acting in the best interest of the chapter, NAACP, or the people she is supposed to represent. She has been accused of taking public stands without the chapter's approval, not having financial records in order, not having records audited per national and local requirements, and running basically a one-woman crusade.

After initial introductions of the people in the room and a welcome, I began my 55 minutes of allotted time with an overview briefing of the Ordnance Corps, which included general topics such as locations, organization, and the initial entry training program model. I then presented a very detailed breakout of gender and ethnicity data.

We then cleared the room of media and went into a detailed discussion of the situation at US Army Ordnance Center and School (USAOC&S). I covered our three main objectives and emphasized that the victims exposed the perpetrators in their statements and that we did not go looking for men to accuse. I stressed that the problem was one of an abuse of power and position. I then went into detail on the ethnicity of victims and subjects.

I commented that the alleged victim ethnicity profile tracked very closely with the population from which it was drawn—58 percent white, 25 percent black, and 17 percent other. To me, it was clear the subjects were "equal opportunity (EO)" sex offenders, and I made that comment to CODEL.

As previously reported and well understood by CODEL, the vast majority of the alleged perpetrators were black (65 percent of all subjects in the multiple ongoing investigations, but all of the drill sergeants [DSs]). I had heard from a retired senior Army general officer (GO) that APG had been assigned a plus-up of black DSs, as requested by a former USAOC&S commander. Blacks made up 48 percent of the staff and faculty of USAOC&S compared with 28 percent for the overall Ordnance Corps.

I then covered the results of legal actions and pointed out the legal system was working fairly—a black accused was not always found guilty.

When I finished, I felt really good I had made the case that the cold, hard numbers showed conclusively that there was no racial bias and that blacks were not being targeted. But then Representative Waters commented that she didn't understand why so many of the accused were black. I was stunned. This was another reminder that there was no way numbers and facts could overcome an emotion-filled agenda, and that was what we were dealing with. This was turning into an instant replay of the December 19, 1996, session with the NAACP.

I played my trump card. The ethnicity of SSG Delmar G. Simpson's alleged victims: Black = 33 percent, Caucasian = 47 percent, and Other = 20 percent. I then said, "Ladies and gentlemen, Delmar Simpson was an EO sex offender. Who in this room wants to go with me to see the parents of the young black female trainee who was victimized and tell them that we are not going to prosecute SSG Simpson because their daughter doesn't count?" Not a word from CODEL.

I then repeated my comment from December 19 to the NAACP: "Ladies and gentlemen, it does none of us in this room any good at all to have all the great black noncommissioned officers in our Army painted as rapists of innocent, young white women. What can we do to keep this from happening? What would you like me to do? I need your help." Again, not a single word was uttered by any of the members of the Caucus or their staffers in attendance.

After my presentation, CODEL met with a focus group consisting of DSs and then with a group of trainees. Following a meeting with the victim support group, CODEL returned to Phillips Army Airfield, where they held a press conference before flying back to Washington, DC. I did not attend the press conference. I waited outside the hanger to say good-bye. Brigadier General (BG) Gil Meyer forwarded me LTC Harkey's executive summary of CODEL's comments at the press conference: "We [Congressional Black Caucus] are more concerned than ever that some-thing is wrong . . . and can the Army investigate this fairly and accurately? We are extremely concerned about coercion and the CID investigative procedures. We are convinced an independent investigation must go forward. Drill sergeant morale is very, very, very low and they want an outside investigation. But, to the Army's credit, these drill sergeants and recruits felt they could come forward and speak openly."

When CODEL exited the hanger, Representative Waters came up to me, took me by the arm, and said softly to me, "General, we really appreciate your sensitivity to this issue." This confirmed in my mind that the CODEL's position was based on a heartfelt need to ensure fairness in the investigations and judicial actions. We all shared this common objective, but it was hard on those of us who were working to make this happen yet constantly pillared in the press as being biased.

I sent an email to the usual cast of recipients with my observations on the day with CODEL. I recommended we stay away from sensing sessions (focus groups) with trainees because they were all new and could only offer anecdotes about what they had heard from others.

The Congressional Black Caucus's allegations got some play in the print media on Tuesday, but not as much as I had expected. In a story in the *Washington Post*, Mr. Mfume took a more measured tone about the race issue. The article quoted him as saying he was not suggesting "race is a singular driving component in this," and that "[w]e really don't know and probably won't know until all the facts play out." The article went on: "Stuart J. Robinson, counsel for the NAACP's Harford branch, said the organization's interests transcend race. 'This is about rank and power and authority and abuse of that power and authority.'" I suspected the NAACP senior leadership's position shift was due to their suspicions of Mrs. Grant's motives after apparently being set up at the media event at her home.

We had one victim, Private (Pvt) Doe 7, who was extremely fragile and apparently suffered from borderline personality disorder. We provided her with maximum care. However, in the opinion of experts, she was not suitable for future duty in our Army. The real tragedy was she had no place to go if she were discharged. We were working a discharge under Chapter 5 (Separation for the Convenience of the

Government) with a secretary of the Army designation for mental/physical health care after discharge to help ensure continued support. BG Mike Kussman, commander of Walter Reed Army Hospital, and I talked about this case often. His folks were extremely helpful, but we would not have been able to do anything without Sara Lister's assistance to get the designation.

BG Daniel A. Doherty called me on March 19 and told me Secretary West had a private meeting with six members of the Congressional Women's Caucus. Secretary West evidently convinced the members an independent investigation would interfere with the ongoing legal actions and could jeopardize the courts-martial. The calls for an independent investigation began to trickle away after this date, but the race issue would continue to ebb and flow.

LTC Black called the next day to report on the Senate Armed Services Committee (SASC) meeting with Senators Thurmond, Levin, Warner, and Snowe. A highlight of this meeting—which I assumed included Secretary West, GEN Reimer, Lieutenant General (LTG) Jerry Bates, and BG Doherty—was that Mr. Mfume was surprised by Janice Grant and the Harford County Branch. This confirmed my earlier information.

Nearly all the meeting attendees agreed that if there were abuses by the investigators, then the judges in the courts-martial could ferret that out; an independent investigation was not needed; and the DAIG would look at the courts-martial proceedings after they were concluded.

Senator Snowe stated that an independent investigation was needed, that it needed to be done soon, and that she was still concerned about APG. She wanted to visit APG.

Secretary West told Senator Snowe he thought the situation at APG happened because Training and Doctrine Command (TRADOC) was reduced too much, too quickly; because one company went bad; and because the Army, as an institution, had noted a change in values of young people.

Major General (MG) Morrie Boyd called me and recommended that when Senator Snowe visited us, I should cover what we did before the November 7, 1996, news conference. Apparently some people in Washington thought we woke up the morning of November 7, found out about the scandal, and reported it on the same day.

On Thursday, March 20, MAJ Susan S. Gibson informed me that Captain (CPT) Derrick A. Robertson had pled guilty in accordance with a plea agreement. He was sentenced to four months confinement and dismissal from the Army. I asked what a "dismissal" was, and she said it was like a dishonorable discharge.

MAJ Gibson also reported that Pvt Doe 7 was very pleased. Two of the primary reasons the government had agreed to the deal with CPT Robertson were (1) to keep Pvt Doe 7's mental records out of public scrutiny; and (2) to avoid putting this very fragile witness on the stand.

CPT Paul Goodwin, who was both the adjutant and my executive officer from January to April, recalled his time as the suicide-watch officer for CPT Robertson:

> Turned out my job was easy since he was in complete denial about how much trouble he was in. I remember him telling me "the Army can't just ignore my ten years of service." I stopped what I was doing and replied. "Actually, Derrick, there are times when your service is ignored. A DWI [driving while intoxicated] is one of those times . . . and having sex with a young recruit when you are their commanding officer is another!" I saw the light bulb go on at that moment.
>
> It still boggles my mind that Derrick could escort Drill Sergeant Simpson to Quantico prison on a Friday and have sex with one of the alleged victims that very Sunday—at his house. Unbelievable. Fortunately his own stupidity was his own undoing . . . He then panicked and had a meeting with [his battalion commander] where he admitted to having sexual relations with the soldier, but that it was consensual.

I received a personal email from GEN Reimer about Senator Snowe's visit scheduled for the following day. His email read:

> Bob, One of the things I told Senator Snowe [during the meeting yesterday with SASC] was that I thought we had cut TRADOC back too far. I mentioned the fact we have cut the chaplains out and I was concerned about the [CASCOM] reorganization we had done. When she visits you she may ask about that. You will want to be prepared, I think, to talk to her about the cuts and the staff you have. I think we need to lay it on the table and if we in fact have cut too much we need to admit that up front. Just a heads-up because I think she's headed your way fairly soon. She asked me after I finished talking to her whether she should go visit the drill sergeants and I told her that I thought that would be okay. I said their morale is down and I would hope that she would tell them that those who are doing what's right are doing a good job. She asked about seeing trainees and I said none of those trainees were there when this took place so they will give you what they had heard. I told her to be sensitive to that, but I encouraged her to come. She is the one who has to be convinced and I think the best way to convince her that we're trying to do right is to let her come over there and see what's going on.

The main point I took from the email was that this visit was a big deal. From the feedback I was getting, Senator Snowe had some strong opinions that would be hard to change. I spent most of the day preparing for the senator's visit and preparing to cover those topics GEN Reimer mentioned. I received calls from several people in the Pentagon rein-forcing GEN Reimer's points.

After the morning update on March 21, I met with Chief Warrant Officer 3 (CW3) Don Hayden and MAJ Gibson to discuss two subjects.

I had had a sinking feeling for a while that I would never get any information from the DAIG on the who-knew-what-when question so I could close out this chapter of USAOC&S history. CW3 Hayden agreed to check all the statements from victims and subjects to see if we could find out why the information on the crimes we had uncovered never reached the right level.

I would later put Johnnie Allen in charge of our internal USAOC&S task force to take CW3 Hayden's input and any other information we had discerned to see if we could answer the question on our own. After reading the victim statements myself, I had a strong feeling we could figure this out even without any external help other than the great CID team.

The second issue was what to do with Pvt Doe 7. Sara Lister's office informed us we would not get a secretary of the Army designation for her. Their guidance was to keep her longer and then do a medical evaluation board (MEB). Our medical team assessed that if we sent her to Walter Reed Army Hospital for an evaluation, she would be sent right back. Because we could not Chapter her out without a designation, and because an MEB was a long shot, we determined we would work to get her into a school, train her for another military specialty, and give her a fresh start in the Army.

Senator Snowe arrived for her visit with Dale Gerry and Peggy Klein of her staff, Senate Staffer Charlie Abell, and LTCs Black and Hawley of the OCLL.

After introducing the senator to the commanders and staff present, I spent about 15 minutes going over the standard orientation briefing on USAOC&S. I then moved into what we titled as a "Special Presentation to Senator Snowe," which covered:

- a timeline of what happened, going back to the spring of 1996 to show her we just didn't wake up on November 7 and say, "Wow! We've got a big problem here!"
- the three main objectives to which we had stayed true since September 13, 1996

- our assessment of what had happened
- the organization and operation of our crisis action team
- the Combined Arms Support Command (CASCOM) reorganization and its negative effect (per GEN Reimer's guidance)
- a list of 17 preventive/corrective actions we took immediately starting in September
- our recommendations for what a commander should do in a situation like this
- six graphs with the same detailed ethnicity and gender data I had presented to the Congressional Black Caucus

The senator asked very good questions concerning the number of cases and the breakout of physical versus nonphysical incidents, the role of chaplains, and the number of soldiers who had been suspended.

Staffer Charlie Abell said, "I have only one question. Who's responsible for this mess?"

I replied, "That's the easiest question you all have asked. I'm the commander, and I'm responsible for everything that happened and failed to happen. What happened was we had some bad people doing some bad things, we found out about it, and we're working the problem. What didn't happen was we did not find out about it soon enough. Russ, tell them why."

Russ Childress then went over again the disastrous effect the CASCOM reorganization had on USAOC&S. Russ had lived through the history, and he provided the senator with much detail, which further emphasized the point GEN Reimer wanted us to make.

I thought that part of the visit went well, with the exception of my interaction with Abell. During my presentation, he asked for some CID data, and I told him I had that information but was not free to release it. He gave me a dirty look. I went up to him at the break before the group went for visits with soldiers, and I told him I would ensure he got the information he wanted. Charlie said very smugly, "General, I can get anything I want from you anytime I want it." This was a classic example of a staff officer wearing his boss's rank.

The group then met with a group of senior enlisted soldiers. This appeared to work better than just DSs, and our noncommissioned officers (NCOs) focused on the effects of the drawdown on training.

The senator and her party then met with our victim support group, who discussed the personnel reductions at the installation level. They had lunch with company commanders and first sergeants, observed gender integrated training being conducted, and talked with female soldiers learning how to repair power generation and water purification equipment. I did not attend the discussions with the leaders and soldiers, so LTC Black wrote up a complete summary:

> Upon returning, Peggy Klein, (Snowe staffer) called to tell me that Sen. Snowe was "very impressed with what she saw and heard from MG Shadley and his soldiers." She "was equally impressed with what has already been done," to prevent such incidents in the future. The Senator may visit other installations in the future, but has no specific plans in mind.

The body of LTC Black's report hit the high spots of what the senator and her team had taken from the meeting:

> His [meaning, my] consistent theme was that they [the command] "did not wake up on Nov 7 and realize they had a problem." They had been working the developing issue all the way, trying to do the right thing.
>
> . . . noted that the CASCOM reorganization resulted in a loss of 886 authorizations (e.g., his [my] immediate HQ staff went from 26 down to 3).
>
> MG Shadley noted that such incidents cannot be completely stopped; some will occur, but we can mitigate the

numbers by enforcing the buddy system, enhanced sexual harassment training, a better trained and more sensitized chain of command, and by adding a training officer to each company. Two keys are education on what is and is not acceptable behavior for a soldier, and enforcement of zero tolerance.

The oft-repeated message from the CSMs and Senior Drill Sergeants was that this "is an NCO issue, not a drill sergeant issue. . . ." It was noted that an NCO would normally be expected to be the first person to see and report what was happening, but "if the NCO is the perpetrator, who is going to do the right thing?"

I later received a thank-you note from Senator Snowe that included, "I found my discussions with you, your company commanders and senior noncommissioned officers both constructive and informative." She concluded, "I appreciate the special Ordnance Coin you presented to me. It was a special pleasure to meet your wife, Ellie, and I hope you will thank her as well for the lovely ordnance potholder."

The Article 15 proceeding in the Beach case was concluded on March 21 by COL Roslyn Glantz, the Garrison commander. He was found innocent of all sexual misconduct–related charges, but guilty of two charges involving improper contact with trainees. This story would be in the media the next few days with the theme being SSG Beach was acquitted of sexual misconduct charges. I did not agree with COL Glantz's finding, but I respected her judgment and thorough review of the case. It also showed each case was being judged on its own merits and there was fairness in the legal system.

On Sunday, March 23, I departed for the National Training Center (NTC) at Fort Irwin, California. The purpose of the trip was to attend the Division XXI Army Warfighting Experiment at the NTC with all the TRADOC commanders/commandants, hosted by GEN Hartzog.

I left for the trip carrying two briefcases full of sworn statements by victims and subjects. I was looking for more clues on who knew what and when. Some of the DSs who were not being charged were expressing concerns about having their "power" diminished. It was clear to me they had known what was going on and had done nothing about it. I was and still remain convinced several DSs knew what was going on, and while they did not participate in the misconduct, they also didn't do the right thing.

We were staying at a hotel in Barstow, California, and we were bused to and from the NTC. On the way back to the hotel one night, the commander of Fort Leonard Wood, Missouri, was sitting in the seat across the aisle from me. He commented loud enough for several to hear that I was stupid for reporting sexual misconduct as rapes. I said maybe so, but it was the right thing to do. His comment really irked me—to put it mildly.

On March 25, the folks back at APG published press release #19, which read:

> Staff Sgt. Wayne Gamble, 36, C Company, 16th Ordnance Battalion, has been charged with desertion (one specification), sodomy (three specifications), assault (one specification), violation of a lawful general regulation (14 specifications of the OC&S regulation governing interaction between permanent party and students), and a general charge including adultery (10 specifications), indecent language (two specifications), and indecent acts (one specification). These charges involve a total of 14 female soldiers and the incidents allegedly occurred between January 1995 and October 1996.
>
> Sergeant 1st Class Ronald Moffett, 30, B Company, 143rd Ordnance Battalion, was charged with cruelty and maltreatment (one specification), violation of a lawful general regulation (four specifications of the OC&S regulation governing interaction between permanent party and students), a general charge including indecent assault (two

specifications), adultery (two specifications), and indecent language (one specification). The charges involve a total of four female soldiers and allegedly occurred between January and December 1995.

The press release also summarized the actions taken so far on the 10 individuals formally charged to date: SSG Beach was referred to Article 15, and that case was under appeal; CPT Robertson was found guilty in a general court-martial; SSG Simpson would have pretrial hearings on March 31; Sergeant First Class (SFC) Brown requested and received a discharge in lieu of court-martial; Sergeant Chestnut requested and received a discharge in lieu of court-martial; and SSG Gamble, SFC Moffett, SSG Gunter, SFC Jones, and SSG Robinson were awaiting court-martial but no dates had yet been set.

SSG Gamble was the one who allegedly had made the comment, "If it worked for OJ [Simpson], it ought to work for us." He was also the one who orchestrated the five female trainees to report to Janice Grant that they wanted to recant their statements.

I had the opportunity to update LTG Bates, the inspector general (TIG) of DAIG, because we were both at the NTC. We talked about who knew what and when, why it happened, the status of the bad actors, and the ethnicity issues. I told him about the CID study done for me of the criminal history of our DSs, with 40 percent having criminal misconduct in their files. I don't think he agreed with me at the time, but he would eventually because his team came to the same conclusion a couple of months later. I later sent him a copy of CW3 Hayden's official report on each of the DSs. While we talked, LTG Bates was very cool to me and seemed in a big hurry to get away from me. When I started talking about the DSs with criminal files, he walked away and wouldn't continue the conversation. I sensed something was wrong.

The announcement of Gamble and Moffett being charged was the topic of an article by Scott Wilson in the March 26 edition of the *Baltimore Sun*. Mrs. Grant was quoted as saying: "It reinforces our belief that this is targeting of black drill sergeants. . . . This is a designed campaign against black Americans. It's a tragedy when you think that this is true in the military."

Friday morning back at APG was the first time in a week that I had met with the team, so we took some time to catch me up. At the update, LTC Gabe Riesco talked about an article Jackie Spinner was working on. He thought it was clear she was working the who-knew-what-when angle. LTC Riesco was also picking up that the secretary of the Army's office and Congress also appeared to be doing their own analysis of the chain of command. LTC Riesco concluded, "Know this is a hard line to take, but we really have sufficient information to hold some of the leadership accountable. Think we'll be called to task on this eventually." I agreed wholeheartedly.

After the update, I sent an email to MG Ken Guest to formally request the return of the AR 15-6 team to USAOC&S. MG Guest responded, "Bob, need to discuss in person." I knew I was in trouble. If the Army would not let me take action, then that meant they were going to punish me.

A story in the Sunday, March 30, edition of the *Washington Post* was sensational, to say the least. The title was "Consensual Sex was Widespread at Army Base—Inquiry Finds Breakdown of Discipline at Aberdeen." Continuing pages had titles: "Misconduct Cases Part of Widespread Problem of Consensual Sex at Aberdeen" and "Lack of Discipline, Oversight at Aberdeen Facilitated Sexual Misconduct, Some Say." To me, this pointed out one of the problems dealing with the media, and it was a prime example of sex sells. The reporters had written a relatively balanced article, but then the editors wrote the sensational titles to get the readers' attention.

BG Doherty and I had a long chat on Monday morning. We agreed this was basically an NCO problem at the SSG/SFC level. He said CID was trying to divert attention away from us, and I told him it would be great if they could do it.

Paul Boyce called and reported that Mrs. Grant was holding another press conference later that evening and had a male DS who was going to make some sort of statement.

At the end of an article in the *Washington Post* the next day, it was reported that at a motion hearing that day, the Simpson defense team wanted to call the Army senior leadership to testify in the case.

There were three paragraphs at the end of the article about the news conference at Mrs. Grant's house the evening before:

> In a bizarre scene today, five men and a woman who said they were sergeants or former instructors at the base appeared at a news conference nearby and alleged in broad terms that the Army was targeting African Americans and ignoring white soldiers. The six covered their heads with pillowcases and shawls and turned their backs to television cameras, they said, because they feared official retaliation if they were identified.
>
> Leroy Warren, Jr., of Silver Spring, a national NAACP board member, accused investigators of using heavy-handed tactics to coerce exaggerated or false statements from white victims against black soldiers.
>
> "In a sign of the times, the KGB has come to America," said Warren who is chairman of an NAACP committee monitoring the Army probe.

I closed out the day with a phone call with COL Tom Leavitt, who headed up the DAIG team doing the investigation for Secretary West. Tom reported that Secretary West had lectured LTG Bates about the DAIG report. I thought maybe this had happened before I saw him at Fort Irwin, and maybe that was why he seemed upset with me and/or the situation. Tom went on to say that Forts Leonard Wood and Jackson were still "different," which meant to me that we would be categorized as not typical of the Army as a whole. We were still the lightning rod for the Army.

Years later, I would learn there may have been more to this confrontation between LTG Bates and Secretary West. I would never find out conclusively all of what went on in the discussions with the TIG and his team with Secretary West, but on March 30, 1998, a friend who was a high-ranking Army GO stopped by my office at my new duty station. We talked about the APG sex scandal and how he thought I was not treated fairly. He asked me if I had talked with LTG Bates, who was then retired. I told my friend I hadn't, and he said I should because he felt LTG Bates would confirm that Secretary West directed him to change the DAIG report to fault me. I sent LTG (Retired) Bates an email and asked if we could talk. He called me on April 14, 1998, and he made the following points, which indicated to me that my friend had the right take on things: "I won't [can't] say the DAIG did a bad job. I talked to a lot of people in the Army about this, and shit happens. Don't lose sleep over this." Whether or not Secretary West directed the report to be changed, I'll never know for sure, but it sure sounded like it to me.

It would take me several years to unravel the Game and the Army's reaction to its Aberdeen sex scandal. And in that time, I did lose a lot of sleep over it.

13

★ ★

THE MOTHER OF ALL TRIALS BEGINS

WHILE IN THE PENTAGON ON APRIL 1, 1997, I STOPPED BY Colonel (COL) Tom Leavitt's office to give him a copy of where we stood on the actions in our get-well plan we had begun to implement in October 1996. I also gave him some charts from Chief Warrant Officer 3 (CW3) Hayden that provided a profile on all the subjects and past offenses by titled drill sergeants (DSs), as well as a large spreadsheet that summarized all the data.

The big news in the Staff Sergeant (SSG) Delmar G. Simpson court-martial motions hearing was the desire by the defense to call senior Army officials to testify and to have charges dismissed because of undue command influence. The statements made by Secretary Togo G. West, Sara Lister, and General (GEN) Dennis J. Reimer in November 1996 were cited as evidence that Simpson could not get a fair trial. The defense was also making an issue of our Crisis Action Team involving Criminal Investigation Command (CID), legal, and public affairs officers (PAOs) by alleging we were interconnected to the Pentagon. The judge would eventually rule against the defense motion.

In no case at Aberdeen Proving Ground (APG) was command influence found to have existed.

We put out a media advisory that identified the 11th cadre member to have charges preferred. SSG Marvin Kelley, 33, a DS in B Company, 16th Ordnance (Ord) Battalion (Bn), was charged with violation of a lawful general regulation, making a false official statement, adultery, and obstruction of justice. The charges involved six trainees and one enlisted member of the staff and faculty.

Prior to the morning update on Friday, April 4, I talked with Major General (MG) Ken Guest on the phone. We had spoken several times about needing to analyze who knew what and when.

MG Guest later forwarded me an email from MG James J. Cravens confirming our discussion about an investigation of the chain of command. MG Cravens said, "DAIG [Office of the Department of the Army Inspector General] is covering this subject in their report to SecArmy this month. When passed to CG [Commanding General], we'll make internal assessment as to adequacy. If insufficient, CG [GEN Hartzog] will launch follow-on investigation. At this point, no action should be taken. Will keep you advised."

The summary court-martial for Pvt Doe 2 began on Saturday and was completed on Sunday. The proceedings and outcome were well covered in the media on Monday, April 7. In Scott Wilson's article in the *Baltimore Sun*, he detailed how the charges of lying to investigators had been dropped, but that she had pled guilty to five offenses not related to sexual misconduct. Wilson went on to write: "After the court-martial, [Pvt Doe 2] shocked local civil rights leaders by declaring that 'race is not an issue' in the sexual misconduct cases where 11 black APG soldiers have been charged since early November." Janice Grant from the Harford National Association for the Advancement of Colored People (NAACP) was quoted in the article as saying, "She [Pvt Doe 2] was a little emotionally confused."

The summary court officer, Captain Chris Eden, provided the following observation a few years after the trial: "Her [civilian] lawyer was being paid for by someone other than herself, and I'm not sure she

Author in Viet Nam, 1971.
(Personal photo)

New Chief of Ordnance and CG, USAOC&S, 1995.
(USAOC&S, used with permission)

Ellie and Remington.
(Personal photo)

Jerry, Cathy, Nathan and
Nicole Stephens. (Personal photo)

TWA Flight 128 Crash Site, 1967. (Courtesy of Kenton County Public Library, Covington. KY)

USAOC&S Headquarters, Simpson Hall. (Personal photo)

Carol Nye and Ellie help cut cake for selection for promotion.

(USAOC&S, used with permission)

Retired General Jimmy D. Ross and Ellie pin on second star.
August 23, 1996 (USAOC&S, used with permission)

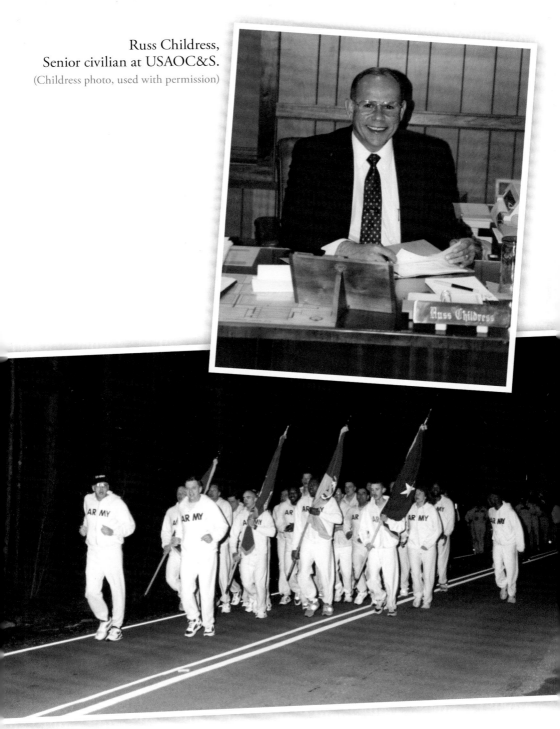

Russ Childress,
Senior civilian at USAOC&S.
(Childress photo, used with permission)

Morning run with Colonel Tom Hooper and 59th Ord Bde.
(USAOMMC&S, used with permission)

Above: Visiting EOD training.
(USAOMMC&S, used with permission)

Left: Visit to map reading class.
(USAOMMC&S, used with permission)

Crisis Action Team. *Front*: Don Hayden, Dennis Webb.
Middle: Charles Vickers, Rob Krauer, Susan Gibson, author.
Back: Pam Royalty, Mary Joe Clark, Johnnie Allen, Gabe Riesco, Cecily David.

(David photo, used with permission)

Ed Starnes, USAOC&S
Public Affairs Officer.
(USAOC&S, used with permission)

Cartoon in Colorado Springs Gazette. (with permission of Charles R. Asay)

COL Webb and MAJ Gibson briefing HNSC. December, 11, 1996.

(USAOC&S, used with permission)

Author and wife Ellie, formal dinner, Fort Riley, KS.

(Personal photo)

Senator Mikulski visit. November 26, 1996. (USAOC&S, used with permission)

Author with COL Roz Glantz and Rep. Bob Ehrlich.
(USAOC&S, used with permission)

CPT Natalie Griffin,
Fort Benjamin Harrison, 1989.

(Griffin photo, used with permission)

Ellie with COL (Dr.) David.

(USAOC&S, used with permission)

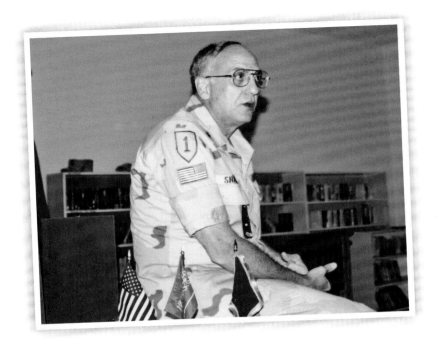

Meeting with Ordnance soldiers and civilians in
Saudi Arabia. (personal photo)

Black Caucus visit, March 17, 1997. (USAOC&S, used with permission)

Senator Snowe visit,
March 21, 1997.

(USAOC&S, used with permission)

Ellie and author with CPT Sheila Bruen. (USAOC&S, used with permission)

Farewell speech,
July 10, 1997.
(USAOC&S, used with permission)

COL John Smith and
BG Gil Meyer.
(Smith photo, used with permission)

GEN Jay Hendrix presenting retirement award. (personal photo)

Retirement reception cake. (personal photo)

Author and Steve Koons as the Blues Brothers at FORSCOM G-4 picnic.
(Personal photo)

Remington and author enjoying sunset. (personal photo)

really talked to him very much. It was interesting that the [civilian] lawyer wrote a letter thanking the government for all the help. I recall at the proceedings that he really didn't understand the UCMJ [Uniform Code of Military Justice] and the government lawyer spent extra time making sure that PVT Doe 2 had a fair trial."

Paul Boyce sent me an article in the April 14 edition of the *Army Times* in which Professor Charles Moskos advocated for "an all-female chain of complaint." There were pros and cons to this idea, but I kept quiet as my experience biased my thoughts on this idea. I had read too many statements where female leaders told female trainees to "Suck it up. That's what we had to do." I was convinced more females in the chain of command may not be the panacea some were looking for.

At our morning update on Monday, April 7, CW3 Hayden was visibly upset by the continuing negative comments the NAACP in general and Mrs. Grant in particular made about his team, especially Mrs. Grant's comments on Channel 45 the night before. The NAACP again reported they had a list of white officers who had not been investigated. Lieutenant Colonel (LTC) Mary Joe Clark (promoted from MAJ) reported that Mrs. Grant told her she was sorry and wanted to apologize to me because this was an old list dating back to December 1996 that had already been examined by CID.

I signed a memorandum through MG Guest and GEN Hartzog to Secretary West. The subject was "Request for DAIG's Report of Investigation—Sexual Harassment Policies and Procedures at Aberdeen Proving Ground." I specifically asked for the report or "at minimum, copies of interviews and other evidentiary matters to determine if we need to take action against any members of the cadre or chain of command." Major (MAJ) Susan S. Gibson drafted the letter to make sure it complied with all regulations and policies. I knew I would never get anything, but I didn't want to be told later that I didn't get the report because I had never asked for it "officially."

As suspected, I never received a reply to my letter to Secretary West. GEN Johnnie E. Wilson came for a visit late in the morning of April

7. I gave him a copy of an email GEN Reimer had sent to GEN Hartzog on April 2, which read, "Bill . . . one of the questions going around is why do we have AIT training at Aberdeen . . . the training base there is pretty small and I don't know if we have looked at consolidating somewhere else or not . . . ur thoughts pls."

This theme of moving Ord training off of APG would continue for several years. The Base Realignment and Closure Commission 2005 would call for the move of Ord training from both APG and Redstone Arsenal, Alabama (RSA), to a consolidated location at Fort Lee, Virginia. After spending over $1 billion in construction, this was accomplished in 2010. Never again would we have a DS sex scandal at APG or any other advanced individual training (AIT) school, for that matter, because in addition to the consolidation move, the Army replaced DSs with platoon sergeants in AIT in 2008.

Later in the day on April 7, SSG Simpson pled guilty to 16 counts of consensual sex or other improper conduct and not guilty to the 78 other counts against him, which included 21 counts of rape. This was not part of a plea deal and was done in front of the judge before the panel (jury) hearing the case was seated. Further pruning of charges the next day resulted in a total of 59 charges, including 19 rapes, that would be heard by the panel.

In the *Early Bird* was an article out of the European *Stars and Stripes* by Karen Blackman: "Two instructors at a military training center [Darmstadt] will stand trial on charges that they sexually assaulted, raped, sodomized, and sexually harassed [22 female] students at the center. A third instructor has been charged in the case but has not yet been ordered to stand trial."

It never ceased to amaze me that with all the publicity and the emphasis on zero tolerance, DSs and other cadre members around the Army would continue to do this stuff.

At a press conference back at APG, the media asked LTC Gabe Riesco if the people working the hotline asked the callers about their race, as alleged by the NAACP. LTC Riesco said no. The personnel taking the call would first determine if the caller needed to talk with a CID agent by using a standard set of questions on a form that most definitely did not include any questions about race. I had not only seen the forms, but also had used them when I worked the hotline on Sunday mornings. Race was not a question we asked.

The process we developed on our own, with no help from the Department of the Army, seemed like a fair system to me. I was personally offended that outsiders would attack the people who staffed the hotline 24 hours a day, 7 days a week, and were forever affected by the stories they heard. They and the CID investigators worked their tails off, and the last thing they needed was criticism.

Captain Paul Goodwin's recollection of his time as the officer-in-charge of the media center and hotline provided a good insight to what our folks did and felt:

> Initially the hours at the Media Center/Hotline were grueling. I worked 36 straight hours the first day . . . got four hours of sleep and came back to work another 18 straight hours. After that I began working 12 hour shifts. I don't recall how long I worked before getting a weekend off.
>
> Many women called with stories of abuse that couldn't be acted upon, but they wanted us to know it had been going on for decades. There was one call that wounded me to hear. A former West Point cadet who claimed she had served at Fort Lee one summer as part of her training said that on her last day they threw a farewell party for her and an Army major offered to drive her back to her room. Instead, he drove out into the surrounding woods and raped her in the car. She claimed this occurred about 15 years prior so she knew nothing could be done about it. Stories such as this were common and affected me greatly. Eight years later [2005] I joined the DOD Inspector General largely due to a need to help make things better in the military.

On April 10, I participated in a conference call from Eglin Air Force Base with the team back at APG, CID, and folks in the Pentagon. We discussed the meeting Secretary West had had with the Congressional Black Caucus earlier in the day. The Office of the Chief of Legislative Liaison representative on the call provided her take on the points the Caucus made at the meeting: (1) racism is widespread; (2) Rep. Waters said the investigation process was not accurate; (3) consensual sex is a problem; (4) victims were pressured to call what happened rape; (5) the caucus wanted to know the race of hotline operators; (6) they were concerned about the recanting of testimony by victims; and (7) all alleged perpetrators should be charged within 30 days.

I also have a cryptic note that indicates Secretary West also met with Erskine Bowles, President Clinton's chief of staff. I thought, "This is great—we now have the office of the president of the United States interested in the APG sex scandal." This was especially ironic because a large number of calls to the hotline were people asking if President Clinton were going to be investigated.

The April 10 edition of the *Washington Times* carried an article on Secretary West's meeting with the Caucus. The last sentence read: "Mr. Cummings said he's learned that the 'No. 1 question' asked of callers was 'are you black or white.'" I had no idea from where or from whom he got such erroneous information.

At the retirement dinner on August 30, 1999, for MG Jim "Chickenman" Wright and his wife, Carol, Representative Norm Sisisky (D-VA) said to me, "I told Togo West not to set up a hotline because it would cause more trouble than it was worth." His caution proved to be spot-on in that the hotline would end up taking on a life of its own. The overall objective for the implementation of the hotline was for a very good reason. But in hindsight, I believe those who pushed for its establishment would now agree they really hadn't considered the possible second and third order of effects.

While out on the range on April 10 observing explosive Ord disposal training at Eglin Air Force Base, I received a call from MG Mike Nardotti (the staff judge advocate of the Army). MG Nardotti said he wanted to tell me why I could not take action at this time with regard to who knew what and when. I scrambled to take notes on his six points:

Secretary West would decide what to do about the chain of command. (This tracked with my understanding that Secretary West had lectured LTG Jerry Bates and also the comment about Secretary West directing the report be changed. This would make a little more sense later in the day when Representative Bob Ehrlich called me.)

Doing anything now would cause big problems later on. (I translated this to mean I was going to be punished later and therefore I could not do anything to any of my subordinates because they could say I punished them in retaliation for whatever happened to me.)

Both he and Tom Taylor from the Office of the General Counsel (Secretary West's lawyers) said to hold off for now.

I couldn't be accused of sitting back, considering all that's been done so far. (I assumed this meant I would not be fired because there were matters in mitigation.)

I should stay on the moral high ground. (I took this to mean that if something happened to me, I shouldn't badmouth Secretary West and GEN Reimer in public.)

If APG was on page 10 of the *Washington Post*, we were doing okay.

I told him we really needed to move out on the who-knew-what-when question to bring all this to closure. He told me, "Don't get out in front of Secretary West on this." I told him I had asked officially for the DAIG report, and then we talked about how the cases were being handled. He seemed satisfied.

I hung up and said to myself, "I'm dead meat."

Representative Ehrlich confirmed that feeling when he told me some people on Capitol Hill "were out to get" me. I asked who, and he said he couldn't tell me, but he gave me some hints. The timing of his call vis-à-vis recent meetings with Congress said a lot to me. He told me to hang in there and that he thought we were doing okay.

On February 15, 1999, an action officer in the Pentagon with access to information told me, "the Black Caucus led the charge to get your head." As with much that had happened before, during, and after the scandal went public, I was sure I would never know all the facts, and I don't know for sure if the Black Caucus was the primary catalyst or not.

It may have been the Women's Caucus that wanted my head, but Susan G. Barnes told me on September 15, 1997, that she polled women in the military after the November 7, 1996, press conference. All said I was a good guy and that nothing bad should happen to me.

It could have been the HNSC Members, Senate Armed Services Committee or staffers.

On September 16, 1997, Paul Boyce told me, "Congress was out for a head." Later that day, GEN Hartzog said to me, "There may have been pressure from the Hill for a flag" (general officers have a flag with the number of stars corresponding to their rank).

On September 22, 1997, Representative Ehrlich's chief of staff said to me, "Letter [I had received a letter of reprimand] was pushed by one or two of the caucus groups. You can guess which ones."

I didn't sleep well that night of April 10, 1997. The calls that day did not bode well for my future, but I had to force myself to think of the big picture and trust what we were going through at APG would make the rest of the Army better. If everything we were going through made our Army better, then what happened to me personally was not all that important in the overall scope of things.

The Simpson trial started on April 11, 1997, so the media would be all over APG. We had organized a public affairs office task force to handle the influx of reporters.

We also published news release #27 announcing that Sergeant First Class (SFC) Tony Cross, 33, had court-martial charges preferred on him with 13 specifications of failing to obey a lawful general order, one specification of sodomy, and three specifications of adultery. The charges involved four alleged victims and took place between February and September 1995.

This case was unique because SFC Cross was the equal opportunity officer in charge of ensuring everyone was treated fairly and equally regardless of race or gender in the 143rd Ord Bn. He was not unique, however, in another way. Cross was another one who apparently was doing the same thing before he came to USAOC&S.

When I was at Fort Riley, Kansas, on August 8, 1997, the garrison command sergeant major told me, "When I saw Tony Cross's picture on TV, I knew he was guilty. I had to counsel him on numerous occasions in Germany for being a married man and constantly hitting on female soldiers." People knew about, at least, SFC Cross and SSG Simpson long before they arrived at APG.

Jackie Spinner of the *Washington Post* was very good, and she figured out the Game. She and Dana Priest wrote an article about it titled "Drill Sergeants Kept Sex Lists, Court is Told," which appeared on the front page of the April 15 edition of the *Washington Post.* They talked about the Game and how the female players were referred to as being "locked in real tight." The article discussed the actions of SSG Simpson and SFCs Cross and Moffett in the 143rd Ord Bn and SSGs Gamble, Robinson, and Kelley in the 16th Ord Bn. Spinner and Priest also reported that "Gamble told the sergeant [an unidentified sergeant at the ordnance school] that he heard the phrase 'locked in real tight' when he was in drill sergeant school in Fort Jackson, SC."

Spinner and Priest said they got their information from defense attorneys and a copy of a sworn statement they had obtained. The *Post* reporters must have had good sources. I knew their claims were true, because I had read the same statement. I had passed all this information on to the DAIG inspectors, but I never found out if they investigated whether Playing the Game was a pattern of misconduct passed on over the years from DS to DS. I suspect they did not.

I doubted these guys were smart enough to come up with the Game on their own. The DS players in the 16th Ord Bn were indoctrinated upon their arrival in 1994 and early 1995 by a senior DS who had since retired. This had been going on for years. As we were leaving our APG farewell dinner the evening of July 8, 1997, a civilian employee came up to me and said, "I'm really sorry you had to go through all of this. It started long before you got here, and it was not your fault." This was another of many instances where I was told this conduct had been going on for years. It appeared many people knew what was going on throughout the Army. It sure would have been nice if someone had told me to watch out for it before I took command.

Some of the stories in the press talked about a breakdown in the chain of command. I called BG Meyer first thing on Monday morning and asked him to make sure the media understood that noncommissioned officers (NCOs) such as SSG Simpson were in the chain of command. In fact, they were the first link in the chain.

LTC Howard Olson, chief of the general officer management office, which reported to the chief of staff of the Army, called to inform me I would be promoted on May 1. All this meant was that I would finally be paid as a two-star general. If the Army senior leadership did not stop my promotion, then I figured I would not get fired. They didn't, and I didn't.

I received a strange phone call from MAJ Gibson. She asked if I had ever been in a taxi with the young trainee (Pvt Doe 8) from Circleville, Ohio—my hometown—with whom I had taken a photo back in August 1995. I said I hadn't. This would become clear in a few days after Pvt Doe 8 completed her testimony in the Simpson trial.

During my initial command assessment in August 1995, I toured training conducted at Edgewood Arsenal, and a young female trainee told me she was from Circleville. Turned out her aunt knew my mom. I took a photo with her and signed it, "Best of luck in your Army career, Bob."

The print and electronic media were filled with testimony from the Simpson trial, and this would continue for days until the trial was over. The reporting was sensational and painted a very poor picture of APG and the whole Army. It got to the point where I would not read the papers or watch TV. I had read most of the statements, and I knew what a bad picture they painted.

For years, the Army had touted the quality of the NCO Corps as the backbone of its strength. The vast majority of the NCOs in the Army were great, hardworking, and honest, and it was a shame a few bad ones were dragging the good ones down.

We received a query from a CNN reporter based on information Mrs. Grant provided him about a sexual misconduct case involving rape and other felonies by a white sergeant major against a black woman. We worked with CID and determined this was a case in New Orleans, Louisiana, in which neither victim nor subject had anything at all to do with APG. This was another instance when the NAACP's making race an issue created a lot of work for us and distracted us from our mission. We felt we needed to look into every allegation. If we didn't, we'd be accused of a cover-up.

Ed Starnes sent his notes from the testimony in the Simpson trial for April 15. In his cover note, he said his notes may be "X-Rated," so open with caution. I noted Pvt Doe 8 was reported to have said to DSs on multiple occasions she would tell "Bob" if she didn't get her way, and she would show them the picture I had took with her and signed "Bob." I also understood she told her chain of command she needed a weekend pass to "go home with Bob." Now I knew why MAJ Gibson had asked me if I had ever been in a cab with Pvt Doe 8. Needless to say, after that I never autographed a photo using anything other than my official signature block.

14

★ ★

WE ARE WALKING POINT FOR
THE ARMY

I RETURNED FROM FORT LEE ON APRIL 16, 1997, AND MET with Private (Pvt) Doe 8 and her uncle from Circleville, Ohio. She had come to apologize for using my first name and for exploiting the fact that her aunt knew my mom. She said she had made a comment about Staff Sergeant (SSG) Delmar G. Simpson to the company first sergeant (1SG), the battalion (Bn) command sergeant major (CSM), and another drill sergeant (DS). She said she also made a statement to Criminal Investigation Command (CID) that SSG Simpson needed to be stopped. I asked her why she hadn't come to see me or had her aunt call my mom. She said she had been taught in basic training to follow the chain of command. Then she said something that really hit to the heart of the problem with these DSs: "Oh, sir. You would not believe the hold that Simpson had over all of us."

The Simpson trial continued to keep the media well fed. An article by Scott Wilson and Tom Bowman in the April 17 edition of the *Baltimore Sun* focused on the chain of command. They reported, "Now some members of Congress are asking how top Aberdeen officers could have failed to know about sexual misconduct which allegedly began in January 1995, that was common knowledge among many of the post's 2,100 trainees. 'I'm concerned with the whole command climate there,' said Rep Tillie Fowler. 'Was Aberdeen an aberration or not? Was there a culture or climate problem?'" Lieutenant Colonel (LTC) Gabe Riesco did a good job of explaining to Wilson and Bowman how we detected problems in July 1996 and began an internal investigation.

I had a good reaction to the article and was personally glad to see some mention that this had started before I got there. When the scandal first broke in the media, I had heard Major General (MG) Jim Monroe told his subordinate commanders at the Industrial Operations Command at Rock Island Arsenal, Illinois, "Aberdeen was straight when I left." I'm sure MG Monroe felt there was no sexual misconduct going on when he was in command. I'm also sure if he had known about it, he would have stopped it. But he didn't know about it. A retired colonel (COL) with close ties to Aberdeen Proving Ground (APG) told me on April 24, 1999, all this was going on during MG Monroe's command tour at APG, but people were too afraid to take any bad news to him.

Major (MAJ) Susan S. Gibson thought Jackie Spinner's article in the *Washington Post* was good. It addressed the rape by fear issue the prosecution was making, and it included the ethnicity breakout of the 10 alleged victims who had testified for the prosecution so far: four white, three black, and three other.

The main emphasis in news reports on April 18 was the judge's ruling that the rape charges would stand in the Simpson cases, even though some of the alleged victims had not protested having sex with him. The judge in effect upheld the argument of constructive force exerted by the DSs because of their position. This also constituted in my mind a formal recognition of the concept of abuse of power.

Bob Infussi from Representative Bob Ehrlich's office spent an hour in the morning with me. Infussi said he had talked with Secretary Togo G. West the day before, and he told Secretary West we were doing a good job. Infussi also said Mr. Mfume, NAACP president, was not concerned race was an issue, but the local chapter (headed by Mrs. Janice Grant) was a problem for Mfume.

I received a quick trial update and then headed to Philadelphia to be the guest speaker at the University of Pennsylvania ROTC Cadet Awards Ceremony. This ROTC unit was one of the five the Department of the Army charged me to help. A week later, I received a very nice note from Roy Exum, the editor of the Chattanooga *News Free Press*:

> General Shadley . . . You may recall meeting me after your remarks last week to the ROTC Awards Ceremony in Philadelphia where my son is a first-year ROTC cadet. I was so moved by your talk I wrote a column, a copy of which you will find attached. Please understand: I was not taking notes nor did I have a tape recorder but I hope the quotes I attributed to you are served as accurately as I can recall. I am mailing a copy of this column to LTC Theodore Majer [the professor of military science at the university]. Thanks so much for what you mean to our country.

The column Mr. Exum wrote contained:

> When Robert Shadley stood at the podium, he apologized for not being much of a public speaker, and went on to say that up until then, public relations hadn't exactly been his long suit.
>
> But life has many twists and turns and suddenly Robert Shadley, a career military man, was having to draw mainly on his principles and values as he stood in the almost blinding glare of intense public scrutiny.
>
> Robert Shadley is a general in the United States Army. More specifically, he is the [commanding] general of the Army's Ordnance Centers and Schools. Last weekend he was in Philadelphia, PA, for the same reason that I was: to attend

the ROTC spring awards ceremony at the University of Pennsylvania.

My mission was simple. All I had to do was help with the cheering.

But, for Gen. Shadley his was more complex. Not only did he have to sit through what must have seemed like his zillionth awards ceremony, he also had to defend the reputation of what is undoubtedly the mightiest and greatest army ever to march on this earth.

Gen. Shadley's command included the Aberdeen Proving Ground, a weapons-testing center located about 30 miles north of Baltimore, MD, and that is where a number of female Army recruits allege they have been sexually molested and raped.

Last week, the nation's biggest newspapers went into great detail about the sordid case, The Washington Post going so far as to include testimony that three of the drill sergeants who have been charged with sexual misconduct had actually been in competition to see who could have sex with the most women.

So when the general stepped to the podium, there was more than a passing interest in what this man would have to say. He began by telling the audience that right now there are over 1 million soldiers defending our country who are good and decent people.

He also said that our nation's judicial system would deal swiftly and promptly with anyone in our society who broke laws and that in the Aberdeen case there would be no exception.

But what bothered me was a story that the general then told of two soldiers under his command, both billeted at Aberdeen, who had come upon an armed robbery in progress not long ago in Maryland.

"We were awfully proud of these men and called the newspapers to see if they wanted to report on their good deeds. Wouldn't you know no reporters showed up."

"A couple of weeks later, when sex was involved, we had over 300 reporters show up the very next morning," he said. As he spoke, I had to wonder if some of us in the news business haven't gotten our values out of line.

Scandal, sensationalism, sex has always been the mainstay of the supermarket tabloids but, as the general pointed out, now the biggest newspapers of our day are publishing such stories on their front pages.

"I haven't had much formal training on how to deal with the press," the general said after the ceremony, "so I just tried to stick with the basic belief of being honest and fair, just like you would leading a group of soldiers."

"Sure we were embarrassed just as any group in today's society would be if something similar occurred, but I get to see many groups like this one." He eyed the nearby ROTC cadets, "and I can assure you our Army is still a proud one."

The general said the same judicial process that he was sworn to defend will deal with any miscreants, just as it has done in the past and will do in the future.

And then Gen. Shadley gave his best line. "Today we honored some of the best and brightest young men and women that this country has ever produced. Some will become officers and leaders in the U.S. Army and that is very exciting for me.

So let's not lose sight of the fact when we have some soldiers step out of line, because ours is the greatest country in the history of the world."

And boy, that's when I got up to help with the cheering.

That column was one bright spot in a sea of bad press, and I was grateful. We did have a great Army, and I never missed an opportunity to tout the service of all the good people who were not getting any publicity.

I called Ed Starnes from the van on the way back to APG from Philadelphia for feedback on the Simpson trial testimony. The company 1SG and the Bn CSM had testified they had done all the mandatory training, conducted anonymous end-of-class reviews with students, conducted sensing sessions, told trainees what was acceptable behavior and what was not, conducted inspections, and investigated complaints. The Bn CSM said Pvt Doe 8's complaint was about being denied a pass, not about sexual misconduct.

I was getting anxious to close out the scandal because I did not want to leave a mess for my successor to clean up. I called and told the team I wanted to meet with Chief Warrant Officer 3 (CW3) Don Hayden, MAJ Susan Gibson, Mr. Robert W. "Rob" Krauer (the deputy installation provost marshal who had become a key member of our crisis action team), COL Dennis M. Webb, and COL Johnnie L. Allen when I got back that evening.

Friday night was a bummer of a time for a meeting, but better than Saturday. We decided upon the following: (1) MAJ Gibson would draft a letter for me telling everyone in the chain of command above us we were going to end the investigation into sexual misconduct; (2) CW3 Hayden would let me know which commanders at other installations I needed to call because they still owed us information on cases his investigators were working; (3) COL Allen would prepare for Lieutenant General (LTG) John E. Miller's visit and investigation of the chain of command; and (4) CID would take over any remaining investigations currently being conducted by the military police.

I met with MAJ Gibson and COL Buzz France on Saturday to discuss the implementation of the actions. We agreed we would establish a local task force consisting of representatives from the Ordnance (Ord) school, the CID, and the installation legal office to take our own internal look at our chain of command. COL Allen (my assistant commandant) would lead the task force and report to me. We also agreed MAJ Gibson would make every effort to get the close-out memorandum to me by next Monday.

I departed Sunday afternoon for Fort Leavenworth, Kansas, to attend the spring Training and Doctrine Command commandant/division commander conference hosted by General (GEN) Dennis J. Reimer. These conferences were for the two-star commanders to meet with the chief of staff of the Army and his staff without our immediate bosses in attendance.

I took the morning update by phone, and then Carol Nye, my administrative assistant, called. She said LTG Jerry Bates was scheduled to brief GEN Reimer on their flight out to Fort Leavenworth later in the day. I hung up with Carol and called a friend in the Pentagon who had firsthand knowledge of what LTG Bates would brief to GEN Reimer. In the conversation, my friend told me, "The civilians are out to get you."

I don't think I ever knew exactly what was going on at the senior levels in the Army and the Department of Defense (DOD). But I had enough friends in just about every organization who notified me when they saw and heard things they did not think were right, especially if they thought I was getting a bad rap.

At the conference ice breaker social, both GEN Reimer and LTG Bates were very standoffish to me. I may have become too sensitive, but I believed in the old saying, "If someone is really following you, you are not paranoid." Too many people were saying the civilians in the Pentagon and on Capitol Hill were out to get me—I had to entertain the thought it was true.

The conference began on Tuesday at 0830. Most of the attendees brought their wives, but Ellie had decided not to attend. Our panel discussion on how we should deal with sexual harassment and gender inequality began at 1030. MG Tom Burnette was in charge per GEN Reimer's direction, and he treated me oddly. MG Burnette was well-connected politically, and I was sure he knew what LTG Bates and GEN Reimer had talked about on the plane out to Leavenworth. The four of us on the panel had 10 minutes each to speak. MG Claudia J. Kennedy kicked off the discussions by talking for over 20 minutes. MG Burdette was next to talk about what he was doing at Fort Drum, New York, to prevent sexual harassment, and he went a little over his allotted time. I was third, and before I started, I was told to keep it short. MG Bob Foley, commanding general (CG) of the Military District of Washington, concluded our session by talking about his "Consideration of Others" program.

In a call back to LTC Riesco at APG, I learned SSG Simpson's company commander testified that he did not know anything about what was going on and that he had told trainees to come see him at any time. This is the same testimony the 1SG and Bn CSM had given the court.

LTC Riesco also reported the defense had rested in the Simpson case and closing arguments would be heard the next day. His take was that the defense witnesses did not say what they were expected to say, and the defense was in a self-destruct mode.

LTC Riesco went on to say he was well into loading sworn statements into a relational database so we could look for trends in leadership failures. I wondered if someone in the Pentagon was doing that for the Army-wide problem.

There was an article in the *St. Louis Post-Dispatch* about a sergeant first class DS from Fort Leonard Wood who posed nude for two female trainees under his supervision. The article went on to say, "He is the fifth drill sergeant from Fort Leonard Wood to stand before a court-martial since the Army sex scandal broke last November. Four more will go on trial in the next two months." Fort Leonard Wood, therefore, had nine DSs in various stages of courts-martial. We had eight DSs in similar situations.

The newspapers and electronic media were filled with Simpson stories as well as the role of women in the military, the concept of rape by fear, and the environment at APG. Final arguments were heard in the Simpson trial, and the case was turned over to the panel for deliberation.

I received an email from Paul Boyce on April 25 saying a total of

1,243 hotline complaints had been forwarded to CID since the hotline began on November 7, 1996. The number of calls involving my units were nearing only 2.5 percent of the total. In another email, Paul reported that in response to a question from reporter Dana Priest, he had passed to her that CID had 325 cases working worldwide at that time. We currently had 12 cases working (not closed out) at APG, which meant our share was 3.7 percent. The way I saw it, no matter how you sliced and diced the numbers, we had less than 4 percent of the alleged sexual misconduct perpetrators in the Army.

The *Washington Post* on April 26 contained an article that caught my eye. The article read: "El Paso-Three Army [commissioned] officers at Fort Bliss have been kicked out of the Army and sent to prison for sexual misconduct, officials said. The three . . . were named in a complaint that accused 19 officers and soldiers at the base of sexual misconduct with enlisted soldiers." The article went on to say these 19 were not in a ring and were in addition to the COL in charge of the Sergeants Major Academy who was accused of sexual misconduct with a civilian. I thought this was incredible—a major (6 months in prison), a CPT (3 months in prison), and a second lieutenant (20 months in prison). Two of the three got more jail time than our former CPT Derrick A. Robertson. Why weren't the East Coast reporters heading to other places like Forts Bliss, Leonard Wood, Jackson, Eustis, and Lee?

We found out one reason why. CW3 Hayden had hinted to Jamie McIntyre of CNN he should head to Fort Leonard Wood, but McIntyre said he couldn't report from there and still make his 1800 deadline in the evening.

A long article appeared in the May 5 edition of *Time* focused on "The Army, already troubled by sexually predatory drill sergeants, has a problem with its recruiters too." It reported on a *Time* investigation of sexual misconduct by noncommissioned officer (NCO) recruiters with potential female recruits who came to them for information prior to joining the Army.

On December 3, 1997 (after I left APG), I had a phone conversation with an old Army War College buddy, MG Mark Hamilton, who commanded the US Army Recruiting Command from July 1997 to August 1998. MG Hamilton told me he had at least one recruiter impropriety case every day. He said, "We have the same thing as APG in Recruiting Command. Just has not hit the press." He agreed with me that while the Army had a great NCO corps overall, we had an NCO leadership problem.

After I departed APG, I spoke with LTC Joe Tedesco, commander of the 553rd Corps Support Bn at Fort Hood, Texas. He told me that while SSG Simpson was in a sister Bn at Fort Hood in the 1994 time frame, he had been doing the same thing he was caught doing at APG. LTC Tedesco said it was common knowledge and no one did anything about it. This confirmed what one of SSG Simpson's fellow soldiers told me after I had left APG.

We published update #22 to the vector report on April 28, emphasizing the following two points in the cover email:

The first point was, "Unless a new and unexpected case comes to our attention, it appears that all cadre/trainee sexual misconduct cases that warrant court-martial are in the court-martial process. In addition, the majority of the Article 15 level cases are completed or with the Garrison Commander for action." I went on to lay out the procedures we would follow for either closing out or forwarding the remaining allegations and investigations of non-felony cases, and I concluded we would implement these procedures unless directed otherwise.

The second point was, "We have formed an internal review and assessment team to assess actions in the chain of command in regard to who knew what, when, and what they did with the information. We will, however, not take action until advised that it is permissible by higher headquarters."

This would end up being like all my updates—no one responded. That was okay, though. I just charged on with what I thought was right, unencumbered by guidance from above.

GEN (Retired) Colin Powell was on CNN's *Larry King Live* on

Monday night and made the following comment about the sex scandals: "But to some extent—not to excuse anything that happened—it's something of a reflection of society from which we drew our troops. Sexuality and these types of activities are much more common and much more encouraged by, you know, the general situation with respect to morals and entertainment and what people seek."

On April 29, Russ Childress reported he had talked further with Bob Infussi in Representative Ehrlich's office about the effect of the Combined Arms Support Command reorganization on the Ord center and schools. Bob was amazed at the depth of the cuts we took, and he wanted to work on an article for the newspapers on this subject. Bob did an analysis and found that many of the allegations of sexual misconduct occurred within six months of reorganization implementation at APG in 1994.

The panel in the Simpson case reached a verdict after 30 hours of deliberation. In addition to the 16 charges to which he had pled guilty prior to the case going before the panel, he was convicted of 47 of the 54 charges that remained after all the motions and legal proceedings. Included in the guilty findings were 18 out of 19 counts of rape. The judge announced the sentencing phase would begin on May 5.

SSG Simpson was sentenced to 25 years in prison, reduction to E-1 (lowest rank in the Army), forfeiture of all pay and allowance, and a dishonorable discharge.

This verdict received extensive coverage in all the media.

After the sentencing, the Army senior leadership appeared to be enamored with the expression, "Aberdeen is an aberration." It took on the connotation that only APG was a problem and the sexual misconduct problem did not exist in the rest of the Army.

In later congressional testimony, the head of Secretary West's panel, MG Steve Siegfried, said he was the one who coined the expression. He said he had meant it only in regard to the number of SSG Simpson's victims and the fact that CPT Robertson was an officer. I wondered why 3 commissioned officers at Fort Bliss getting prison time was also not an aberration, or why the NCOs at other installations who had 5 to 10 victims were also not aberrations. What number of victims per perpetrator did it take to be an aberration?

As MG Siegfried told me a few months later, "The Army couldn't fire you. If they did, they would have to fire a lot of installation commanders." It appeared to me GEN William W. Hartzog, GEN Reimer, and Secretary West would also have had to have been fired if the problem were indeed Army-wide. To me, "Aberdeen was an aberration" was a convenient battle cry for some to protect the Army as an institution. It also seemed to me the DOD was content with the Army being the point service. What it boiled down to was, it was in everyone's best interest, except ours, to have APG painted as an aberration.

I talked with COL Tom Leavitt, and he told me LTG Bates was going to brief GEN Hartzog on May 5 on the DAIG team's findings regarding the APG situation. He added, "Hartzog would not like what they had to tell him." I took this to mean bad things would be said either about all the training bases under his command or about me personally, as it was common knowledge that GEN Hartzog and I were friends.

On May 1, 1997, there was an editorial in the *Washington Post* titled, "Women in the Military." Paul Boyce alerted me to the last paragraph, which did not bode well for us in leadership positions at USAOC&S. However, I could not argue with the last sentence:

> In the end, what happened at Aberdeen strikes us less as a case of hormones inevitably running rampant than as a blatant failure of command. Where were the high-ranking officers while the accused drill sergeants organized their sex ring? They obviously were not communicating the Army's admirable policies on sexual harassment and crime, they weren't making clear that violating those policies would break your career, and they didn't know, or want to know, what was going on at the base. They, too, should be held accountable. The Army has an obligation—and, we have to believe, the

ability—to make sure women in its forces work in a noncombat zone when they are working and training on the base.

MG Mike Dodson, CG of the 1st Infantry Division (Mechanized) at Fort Riley, Kansas, sent me an American Medical Association survey, "Strategies for the Treatment and Prevention of Sexual Assault." Mike was an old friend, and we had served as colonel commanders together in Desert Shield/Storm. In his cover note, he said, "Think the task we have ahead is more difficult than we think. Take a look at the feelings of young people on the page with the yellow tab."

A survey of boys and girls ages 11 to 14 showed:

- 51% of boys and girls said forced sex was acceptable if the boy "spent a lot of money" on the girl;
- 31% of boys and 32% of girls said it was acceptable for a man to rape a woman with past sexual experience;
- 87% of boys and 79% of girls said sexual assault was acceptable if the man and woman were married;
- 65% of boys and 47% of girls said it was acceptable for a boy to rape a girl if they had been dating for more than six months.

If we thought things were tough now with new recruits, it appeared the future would be even more difficult. A COL who reviewed sexual misconduct at several Army installations and had read a lot of statements by victims told me he was amazed at the promiscuity of today's young male and female soldiers—and not just at APG. These data points supported GEN Powell's observation on *Larry King Live*.

In a related area, I sent an email to GEN Hartzog and several others about the need for a command surgeon at the USAOC&S to help us identify trainees whose life experiences made them susceptible to abuse.

I cited a study by Leora N. Rosen, PhD, of the Department of Military Psychiatry at the Walter Reed Institute of Health in Washington,

DC. It reported: "48% of female soldiers and 50% of male soldiers met criteria for childhood physical abuse; 51% of female soldiers and 17% of males met at least one of the two sets of criteria of childhood sexual abuse; and these soldiers are subject to victimization."

While I was not trained in this area, it certainly appeared obvious to me that at least some of the female trainees who played the Game had grown up having to give sex to adults just to get by, and they felt that's what they had to do in the Army.

As I read through the statements from the alleged victims and subjects, it struck me that the NCO players claimed they were absolutely sure they could pick out the trainees who were willing to play. My personal assessment was that they probably weren't that precise in their selections. At a minimum, they had sexually harassed the females not labeled as willing participants. And in many cases, those trainees still ended up in the NCOs' Game because of fear, past childhood abuse, or wanting to appear cool to their female friends who played.

On May 2, COL Allen and the team reported back on their analysis of culpability in the 143rd Ord Bn at Edgewood Arsenal, Maryland, in not responding to actual or perceived sexual misconduct. I thought the team did a great job of extracting the information they needed from existing documents. The assessment validated my desire to move quickly regarding any leadership failures. Two leaders who I thought warranted discipline had either departed APG or would depart soon.

We had a quiet weekend, but one article by Courtland Milloy in the Sunday Metro Section of the *Washington Post* caught my eye. Milloy addressed the race issue in the APG cases head on. After addressing the calls by the National Association for the Advancement of Colored People (NAACP) for an independent investigation, Milloy stated:

> But which is really the more pressing concern at this time: How the Army treats black drill sergeants allegedly caught with their pants down? Or, how black people treat one another, starting with husbands and wives—the heart and

soul of the family and the key to the advancement of black people in the United States? . . . At a dramatic, tearful news conference arranged by Kweisi Mfume, president of the national NAACP, some of the white female trainees seemed to be saying that they had not been raped but rather had engaged in consensual sex with the black sergeants. How could the NAACP regard such "confessions" to be a victory for black people? Even by the women's revised accounts, the alleged actions of the black soldiers still constituted an offense against the soldiers' wives. Grant says she thinks the reason black drill sergeants have been targeted is because the military is downsizing and looking for an excuse to get rid of some blacks. . . . The bottom line is that most of the black drill sergeants charged with rape, like Simpson, are married. And while concerns about racial discrimination in the investigations are legitimate, the fact remains that if these men had been faithful, none of them would have been in the mess.

15

★ ★

THE ARMY STRUGGLES WITH
ACCOUNTABILITY

I N AN ARTICLE IN THE MAY 5 EDITION OF THE *Baltimore Sun*, I was glad to see Scott Wilson talk about the Army needing to do a better job of selecting drill sergeants (DSs). He pointed out that Staff Sergeant (SSG) Delmar G. Simpson had two incidents of misconduct in his personal history before becoming a DS. I was hopeful that maybe something would be done about DS selection and the other issues we had been raising for several months.

Ed Starnes called and told me the public affairs office at the Training and Doctrine Command (TRADOC) had informed him the report from the Office of the Department of the Army Inspector General (DAIG) was at TRADOC headquarters and the staff could read it, but not make a copy. This really upset me. I was the one who really needed to know that information, not a bunch of staff officers two levels above me.

General (GEN) William W. Hartzog called that evening and summarized his session with the DAIG. I hurriedly tried to catch every word and write it down.

He said they had told him nothing he didn't already know about Aberdeen Proving Ground (APG), and their report was what he had expected. It was written as though we had no resources at APG (at least we had gotten that point across). According to DAIG, my main failing was that I ignored the training base because of other things. He said the question was: Could I continue to do the right thing? That is, did they believe me or not? (This was personally and professionally insulting, but I kept my mouth shut and continued to take notes.) Could I continue to do the right thing to ensure corrective actions were implemented?

GEN Hartzog confirmed that sexual harassment was the only thing covered in the DAIG report.

He went on to say he wished I had done some things differently. He was concerned the sergeants major weren't doing their job. He planned to write General Ronald H. Griffith, the vice chief of the Army (VCSA), to get a decision about my future. (The VCSA was the enforcer in the Army for generals.) He said I'd done nothing criminal.

When I hung up, it was another one of those times when I said to myself, "I'm dead meat."

I called MG Guest. His advice was, serve as a two-star for two or three years, then bail out. But see it through—don't make it easy on them by retiring now.

I told Ellie about my call with GEN Hartzog and let her know things did not look good for me personally. She told me in no uncertain terms what she thought of the Army. I felt obligated to see all the corrective actions through and give my best effort for my remaining time in the Army. First Sergeant (1SG) Ernie Knight, who had served as a 1SG for me as a company-level commander in two units, would say, "Sir, they can shoot you, but they won't eat you."

Reporters Spinner and Priest continued the race theme in an article in the *Washington Post*. Their article stated: "A great deal of pressure was brought to bear on the females to make the case the military wanted to make, said Rep Earl F. Hilliard (D-AL), first vice chairman of the congressional Black Caucus. These women were forced to lie in many instances." Hilliard was quoted as saying, "This is just the beginning of the judicial process. But when you have a case like this situation, where you single out people because of their race, we can't sit back and let it happen." I would later learn the judge advocate general of the Army did tell prosecutors Army-wide to charge rape if the sex was not totally consensual. As with many aspects of this scandal, it would take time to unravel this part.

Chief Warrant Officer 3 (CW3) Don Hayden brought by the final version of his survey of misconduct by DSs who had served or were still serving at US Army Ordnance Center and School. I sent a copy to Lieutenant General (LTG) Jerry Bates, the inspector general (TIG), and to LTG Frederick V. Vollrath, the deputy chief of staff for personnel at the Department of the Army. I reminded them I was sending it as I had promised in April during the winter TRADOC commandants/division commanders' conference at Fort Leavenworth.

Paul Boyce called and reported that the announcement of court-martial charges against Sergeant Major of the Army (SMA) Gene C. McKinney would be preferred later in the day. We received the press release stating he was charged with "maltreatment of soldiers (four specifications), assault (two specifications), adultery (one specification), communicating a threat (two specifications), indecent assault (four specifications), obstruction of justice (two specifications), and solicitation of adultery (three specifications). The charges and specifications involve three female soldiers and one female sailor and allegedly occurred between Oct 94 and Mar 97."

I was still seething about the DAIG report being circulated around TRADOC headquarters and our not having a copy. I sent MG James J. Cravens another email asking for a copy to be express-mailed to me. I never received it.

We held a farewell luncheon for Captain (CPT) Jerry D. Stephens, who was scheduled to depart soon for a class at Fort Leavenworth and then an assignment in the Pentagon. CPT Stephens had done a marvelous job. He and his wife, Cathy, and their children, Nathan and Nicole, were part of our family. I would miss him deeply for what he did to make my job easy in a very tough time at APG. The good news was that CPT Stephens got to see how the Army operated at higher levels—but the bad news was that CPT Stephens got to see how the Army operated at higher levels. Like any large organization, the higher up the chain, the more politics come into play.

I selected CPT Sheila Bruen to be my aide when CPT Stephens left. CPT Bruen was a superb officer who would fit in well with the team.

I called GEN Hartzog's aide, Lieutenant Colonel (LTC) Dave Stahl, and talked with him about the DAIG report. I figured he had attended the May 5 briefing. In any case, good aides always knew what was going on. I told LTC Stahl his boss needed to understand this was a bigger problem than just APG. As LTC Stahl explained it, GEN Hartzog knew there were problems elsewhere, but that he thought they were old cases. He agreed with me that the Army senior leadership was keeping GEN Hartzog in the dark. I assumed this was either because they might discipline him or because they knew he and I were friends. LTC Stahl concluded our conversation by saying GEN Hartzog really supported me, but he had to write the VCSA to get me approved to process the rest of the actions.

In retrospect, not being political worked to my disadvantage. I didn't want to appear self-serving or suggest I was not taking responsibility for everything that had happened or failed to happen, so I didn't tell our story to the senior leadership as I should have. I had clear access to all of the uniformed military in key leadership positions, but I did not take advantage of that access.

The most telling and explosive article in the *Early Bird* on May 8 was an Associated Press article from the *Washington Times*, "Witness: Sergeants planned trainee sex," about the DS Holloway trial at Fort Leonard Wood. The DSs had the Game going at Fort Leonard Wood exactly as in the battalions (Bns) under my command. The article stated:

A 14-year Army veteran being tried on charges of abusing trainees was part of a group of drill sergeants who cooperated with each other in arranging to have sexual relations with trainees, one of the sergeants said. However, he said at least some of the arrangements were made without the trainee's knowledge.

Sgt. Ralph Quander, who cut a deal with prosecutors in exchange for his testifying against his fellow drills sergeants, testified yesterday that Sgt. Holloway told him they needed to watch out for each other. "Sergeant Holloway stated that, if I was interested in a certain private, to let him know so we didn't bump heads," said Quander.

Sgt Quander, 28, said the drill sergeants had a system of sneaking recruits out of their barracks for dates, even arranging to have trainees on guard duty to be moved away from doorways (so trainees and drill sergeants could come and go as they pleased), and then covering for each other if they were asked about their activity. And they knew to stay away from privates who were already involved.

There were too many similarities between activity at Fort Leonard Wood and APG not to send up the red flag that this gang behavior was probably a problem at every installation where male DSs supervised female trainees. It also reinforced the comments I read in some of our soldiers' statements that the DSs learned this in "after-hours instruction" at the DS School at Fort Jackson, where MG Steve Siegfried, the head of the secretary of the Army's panel, had been the commanding general from December 1991 to March 1994.

I was surprised to read an article in the May 9 edition of the *Washington Times* by Rowan Scarborough about Representative Roscoe Bartlett (R-MD) introducing a bill in the House, backed by 81 other members, that directed the military to end gender-integrated initial entry training. The article referenced court testimony of misconduct at APG and Fort Leonard Wood. Mr. Bartlett cited the sex-club atmosphere as a reason to end mixed-sex training at the boot camp level and put female DSs in charge of female recruits.

Bartlett's action would amount to a declaration of war to the women's groups who favored gender-integrated training. Representative Jane Harman (D-CA) was quoted in the Scarborough piece as saying, "If we roll back, then we are punishing women twice. We are the victims of the sexual abuse, and secondly, we lose our opportunity to succeed in roles in the military."

LTG Thomas G. Rhame called and advised me, as MG Guest had, to hang in there for a couple of years, keep my mind healthy, and then retire as a two-star. He said he and Lin, his wife, had been concerned about me. More than three years later on May 10, 2000, LTG Rhame called to check on me. He said that when all this at APG had started, GEN Dennis J. Reimer had told him, "Bob Shadley will be just fine." But a few months later, GEN Reimer just shook his head when LTG Rhame told him, "Bob Shadley is not okay."

Sex, fraternization, race, and gender-integrated training continued to receive media attention. The SMA McKinney fallout was also affecting us. Race was being raised as an issue in his case, and every time sexual misconduct and race were discussed, something about APG also appeared in the media.

I talked with my lawyer, Major (MAJ) Susan S. Gibson, the first thing in the morning on May 13. She informed that SSG Wayne A. Gamble wanted to deal and wanted to provide testimony against the other DSs, including SSG Vernell Robinson. I also found out SFC Cross wanted to make a deal. These two would really help us understand the Game and the culpability of the chain of command. MAJ Gibson estimated the court-martial cases for our DSs would run through August 1997. The accountability issue was also heating up at higher levels because people were beginning to ask why GEN Reimer was not aware of his own SMA's (McKinney's) alleged misconduct.

Colonel (COL) John A. Smith called to let me know MG Siegfried and the panel would be visiting APG in the near future after all. I assumed MG Siegfried was coming just so he could say he had been to see us.

The morning update on May 14 was COL Cecily David's last meeting. She had been reassigned to Fort McPherson to be the US Army Forces Command (FORSCOM) command surgeon. Ellie and I would end up living just a few doors down from COL David and her husband, Winston, on Staff Row. We would miss her at APG. Her focus on caring for people and her levelheaded approach were exactly what we needed. She and MAJ Pam Royalty made a dynamite team covering mental and physical health issues.

GEN Reimer sent out an email on "Army Values" that stated, "Values are at the core of everything our Army is and does." This email announced that "Character XXI is part of a Total Army program designed to teach and reinforce Army values." This was a super idea, and I hoped this in effect would accomplish the same thing we were pushing for—a noncommissioned officer (NCO) Code of Conduct. On May 19, I forwarded this email to my subordinate commanders and covered it in meetings with DSs, company commanders, and instructors to again stress the importance of setting and enforcing the highest ethical and moral standards.

In an office call with Chaplain (Colonel) Kalyanapu, the new APG installation chaplain, I learned the USAOC&S had been shut down for a couple days a few years prior due to an investigation of sexual misconduct allegations. This must have been kept quiet because this was a revelation even to me. I continued to be shocked by the fact that this sexual misconduct had been going on for years and no one had attacked this problem and sought a fix.

At the Asian-Pacific American Heritage luncheon, the guest speaker paraphrased the words of Reverend Martin Luther King, Jr., in his famous "I have a dream" speech by saying, "Judge not by the color of skin, but content of character." I was sitting with Mrs. Grant, and she told me she really objected to the definition of rape as used in the Simpson trial. I suggested, again, that we work together to help improve both the image of the African-American male and the whole APG community. Unfortunately, this had not happened and never would.

COL Allen confirmed that MG Siegfried and his panel were coming to see us. I called TRADOC headquarters to find out why. I later learned from another source that a staff principal in the Pentagon had recommended MG Siegfried go to APG for "image." This confirmed my own assessment and must have been true because we were given one week's notice of the panel's visit when other installations received 30 days' notice.

I assumed the panel would visit only us at the USAOC&S, even though there were more than 50 other organizations on APG and they had more hotline calls than we did. The panel visit was a punch-the-ticket event and would obviously be negative. The panel's report, I later learned, "would be written through the eyes of Aberdeen." Less than 4 percent of the Army's problem would be the eyes for the entire US Army.

On May 15, I sent MG Guest an email letting him know we had finished our assessment of the chain of command and were standing by to take action. My last paragraph read, "Would appreciate expeditious guidance on what higher headquarters wants done. Have officers and NCOs who are in limbo. I'm holding efficiency reports and award recommendations, but allowing soldiers to PCS [move on to their next duty station]. If I flag them [stop all favorable personnel actions], I must say why. If I say why, I'm out in front of the Secretary of the Army which, per TJAG [the judge advocate general], I am not to do."

I departed for the Naval War College at Newport, Rhode Island, to attend the Joint Ordnance Wargame '97 on the morning of May 16. In calls back to APG, I learned from both COL Buzz France and CW3 Hayden that SSG Gamble was "really singing." He was providing details on how the Game was played, who the DS and female players were, who in the chain of command knew and did nothing about it, and specific details to solidify the cases against DSs Robinson, Cross, and Kelley.

It appeared that DSs Gamble, Kelley, and Robinson were a gang and at least one other DS knew and did nothing about it. We also concluded the company commander knew or was dumb. CW3 Hayden and his team would look to see what the bridge was between the DSs in the two Bns, even though they were separated by 10 miles. I then talked with MAJ Gibson, who told me SSG Robinson also wanted to make a deal.

In a May 18 *Washington Post* story about the accountability of the chain of command, Patrick B. Pexton was spot-on. He wrote:

> And then there is Aberdeen. Now, no one can suggest at Aberdeen the top Army officers have condoned or engaged in the horrible misconduct alleged of the drill sergeants. The blame should land squarely on those drill sergeants who had any kind of sexual contact with their recruits, consensual or not. Drill sergeants all know such relationships between leader and led are not only illegal but immoral. And drill sergeants who say their charges solicited sex have forgotten their duty to just say no.

GEN Hartzog called to tell me that for my next assignment, as part of a normal rotation and not a firing, I would be going to FORSCOM as the director of logistics at Fort McPherson in Atlanta, Georgia. When he asked if I were okay with the assignment, I said I was. He said there was no problem with my staying in command at APG until my normal two-year command tour was up in July, because any lawyer could prove I had done nothing criminal. He told me to keep running the legal stuff through MG Longhouser and to send him my recommendations regarding the chain of command.

I understood from a friend in the Pentagon that Secretary of Defense William S. Cohen was letting each Service do its own thing in combating sexual harassment and misconduct. Just as the Army senior leaders were content with APG being the point for the Army, it appeared the senior leaders at the Department of Defense (DOD) were content to let the Army take the lead for the DOD.

At the weekly morning update on May 21, CW3 Hayden reported SSG Gamble had provided additional details of the Game in another interview two days prior. His statements provided very useful insights into figuring out what had happened.

MG Siegfried, Brigadier General (BG) Evelyn P. Foote, and nine other members of the secretary of the Army's Senior Review Panel on Sexual Harassment arrived on May 21.

Our command team met with MG Siegfried in our command conference room while BG Foote and the nine others headed out to meet with DSs, students, and instructors.

MG Siegfried pontificated for almost an hour about how his panel was looking at the human relations environment and focusing on dignity and respect. He stated his findings would not be reported by installation, but would be an Army-wide roll-up.

I then gave MG Siegfried a detailed briefing that included 7 charts with all 40 lessons-learned we had compiled. I also covered the ethnicity data, timeline of events, our objectives, our analysis of what happened, and our assessment of how this happened.

MG Siegfried appeared not to be interested, but I gave him a copy of my briefing, the May 13 edition of our vector report, the results of the Defense Equal Opportunity Management Institute's survey conducted in the summer of 1996, and CW3 Hayden's analysis of the misconduct in the records of DSs.

We than gave MG Siegfried a tour of APG and Edgewood Arsenal to see the training facilities and the barracks areas. As I recall, I was with him the whole time, and he did not personally meet with any soldier groups. He never left my sight. The panel members departed via helicopter at 1500 hours.

MAJ Gibson called and said she had spent an hour with MG Siegfried's team and got the distinct impression I was being set up for something. I strongly suspected MG Siegfried and his team would leave APG with a list of nothing but all that was wrong. Because both the panel and TIG worked for Secretary West, it was not inconceivable to me that he would orchestrate the results to fit his agenda. It was appearing more and more as though one of Secretary West's agenda items was to discipline a GO to show Congress and the Office of the Secretary of Defense that he was taking firm action. This would certainly satisfy the "politicos" who were calling for a head and also make him look good as he lobbied for a cabinet position in the Clinton administration.

The feedback I got from our people who had sat in on the sessions with the Siegfried team was that our soldiers were relatively positive about what we were doing. All the DSs said they would still be DSs if they had to do it all over again. A couple voiced a concern that due to the scrutiny, their power had been diminished, but another answered that he did not feel stripped of his power.

COL Allen called LTC Linda Thompson from Sara Lister's office. LTC Thompson was on the panel and had participated in the sensing sessions. COL Allen's email stated: "She noted nothing came to her attention which required the attention of the chain of command. She has had little time for any discussions with MG Siegfried or BG Foote, but believes they have no issues or we would have been informed."

MAJ Gibson reported the letter with my recommendations on disciplinary action for members of the chain of command was ready for me to sign and forward on to GEN Hartzog. I was still not quite sure what would happen or who would take action, but I felt it was my job to force the issue of accountability and at least start the ball rolling.

GEN Hartzog called to say he had talked with MG Siegfried and was told the panel's report would indeed "be seen through the eyes of Aberdeen." This confirmed what MAJ Gibson had told me about my being set up. The Army senior leadership would be able to say APG was an aberration and that the focus of the Army should be on the prevention of sexual harassment, because that was the "real" problem. Not only were we doomed to be the scapegoat, but the Army was doomed to fail in ever solving the problem of felony sexual misconduct.

Later, I received a call saying GEN Hartzog wanted to visit on Wednesday, May 28. I learned MG Siegfried had told him he needed to get up to APG as soon as possible because the DSs had told the panel things were really bad and GEN Hartzog needed to personally see for himself. This did not track with what my officer described in her after-action report of the sensing sessions. It also did not track with LTC Thompson's comment that nothing had surfaced that required the attention of the chain of command. It did track with MAJ Gibson's sensing, however.

It was becoming clear to me that my head would be the one rolling.

16

★ ★

ABERDEEN IS NOT THE FIRST
SEX SCANDAL

I WAS DUE TO LEAVE FOR FORT LEE FOR TWO DAYS ON MAY 26, 1997. Colonel (COL) Johnnie L. Allen took on the task of coordinating General (GEN) William W. Hartzog's visit on May 28. I flew back to Aberdeen Proving Ground (APG) from Fort Lee and was on hand to meet GEN Hartzog upon his arrival. He and I had a private meeting in my office for over an hour. I provided him the same briefing I had presented to all our other visitors, but I did not go into detail on any specific case of alleged drill sergeant (DS) misconduct, as the lawyers advised. I gave him several documents, which included a copy of the memorandum I had sent to Lieutenant Generals (LTGs) Frederick V. Vollrath and Jerry Bates detailing the misconduct in the DSs' files before they came to APG. I also presented him a thorough explanation and justification for recommended actions against six officers and senior noncommissioned officers (NCOs) in the 143rd and 16th Ordnance (Ord) Battalions (Bns) who did not take action to stop and correct sexual misconduct.

We chatted after I briefed him. He said we could not send actions to

189

MG John M. Longhouser. I knew why but didn't say anything. That was an issue still developing. GEN Hartzog called MG Mike Nardotti (the judge advocate general of the Army), and MG Nardotti told him I should instead send my recommendations to LTG John E. Miller. All this delay was frustrating. My frustration turned out to be well founded because I later learned that one officer, who I strongly felt deserved some form of adverse action, was eventually selected for early promotion to the next higher grade.

On a personal note, GEN Hartzog said I would not be promoted and would not receive a Distinguished Service Medal (the Army's highest peacetime award) when I left APG. He went on to say the good news was that no one wanted to relieve (fire) me. He concluded by saying I had to take the actions to clean up the US Army Ordnance Center and School (USAOC&S) and then let someone else do the healing. None of this was a surprise to me.

I then took him on a tour of the school's facilities, and we observed tank and artillery mechanic technical training before he met with currently serving DSs. Afterward, he gave me a list of the DSs' comments, which he collectively described as "whining over little things."

The *Washington Times* on May 29 predicted that US Air Force (USAF) GEN Joseph Ralston would be announced as President Clinton's nomination to be the next chairman of the Joint Chiefs of Staff (CJCS) to replace Army GEN John M. Shalikashvili. This would later turn out to be problematic.

Other media stories included one by Jackie Spinner, who noted that in the DS Vernell Robinson trial female trainees gave testimony about the Game and how Robinson tried to get them to lie.

That evening, the court-martial panel announced their findings in the Robinson trial: guilty on 19 counts and innocent on 1 count—wrongful interference with the US mail. Staff Sergeant (SSG) Wayne A. Gamble's testimony was limited and did not include the National Association for the Advancement of Colored People (NAACP). After months of scrutiny and accusations regarding the race issue, the trial counsels (prosecutors) should have focused more on how the DSs had used the NAACP as attempt to get off. While it may not have been directly important in the Robinson case, it was important in the court of public opinion and to the US Army's reputation overall.

In a story about the Robinson verdict, Jackie Spinner reported details of the Game SSG Gamble had testified about to the court as part of his plea deal with the prosecution. Robinson was subsequently sentenced to six months' confinement, reduction to the lowest pay grade (E-1), forfeiture of all pay and allowances, and a dishonorable discharge.

I had received letters, emails, and calls of support almost every day since November 7, 1996, but the one I received Friday afternoon from MG Ken Guest's administrative assistant, Dotti Tilmon, touched me. It summarized just about all the messages of support I had received and would continue to receive:

> Our "boss" let me in on a little secret, and that is that you are going to FORSCOM [US Army Forces Command at Fort McPherson, Georgia]. I was so afraid you would decide to retire and I know that you have much more to give to the Army. I can't begin to imagine how difficult these past several months have been for you and your family.
>
> For that matter, how you maintained your sanity throughout it all—the media, in particular, questions from the "Hill," demands of four-star generals, etc.—and yet you managed to do your job always with a smile on your face (or at least every time I have seen you) remains a mystery. (Look at it this way. How many two-star generals does the president know by name?) You are a remarkable person and I hope you will continue to come and see us after you get to Atlanta. Good luck. And thanks for hanging in there. The Army still needs compassionate, caring leaders.

June started out as May ended—the media was filled with all the sex scandals in the Army, Navy, and Air Force; the debate for and against gender-integrated training getting a lot of attention; and race was still an issue in some cases.

In an article by Sean Naylor in the June 9 edition of the *Army Times*, GEN Hartzog was reported as saying that within the next two weeks, he would decide on who in the chain of command at APG would be held responsible. He was then quoted, "Nothing has changed my mind that there were a few bad apples [at Aberdeen]. I don't believe—and there is no indication [from] any of the inspections that are going on—that basic training . . . is broken in any way."

Both Chief Warrant Officer 3 (CW3) Don Hayden and Major (MAJ) Susan Gibson told me they would be reassigned soon because we had completed all the investigations and initiated all the appropriate legal proceedings. I wanted them to stay until I left, but it would not have been fair to them or their families, who had really sacrificed to support them.

In a conversation with Brigadier General (BG) Daniel A. Doherty, who led the criminal investigation, he told me he was going to be investigated. The investigation would address allegations of racial bias by the Criminal Investigation Command (CID) investigators and his granting of immunity to female trainees so they would talk without fear of being held liable for their actions. In retrospect, if the Army had spent more time investigating the "bad guys" and less time on the "good guys," maybe something positive would have come out of all of this.

Ed Starnes, my civilian public affairs officer (PAO), called and said the *New York Times* wanted him to confirm a general officer (GO) at APG was being relieved for adultery. Ed told the caller it was not me and that I would be leaving on a normal reassignment. BG Gil Meyer at the Office of the Secretary of the Army said the GO in question was MG John Longhouser.

I called John to let him know how sorry I was. I felt I had played a part in the destruction of his career. If the scandal had not evolved as it had, there would have been no hotline. I was there when the anonymous call came in about an alleged affair he had had with a civilian woman while he and his wife were separated more than five years ago. John told me he was going to retire. This was one of the lowest points in my military career.

I found an interesting article by Vince Crawley and Tonja Arch in the June 3 *Stars and Stripes,* and it would prove even more interesting later. It was about the trial of Sergeant First Class (SFC) Julius Davis involving 35 charges of misconduct against females at the training center in Darmstadt, Germany. Crawley and Arch reported that SFC Davis was also identified in a hotline call from a female caller who "claimed that in 1993 she was twice raped by SFC Davis when he was a drill sergeant at Fort Benjamin Harrison, IN."

As I was beginning my post-retirement job search in 1999, my ethics advisor was Lieutenant Colonel (LTC) Natalie Griffin of the FORSCOM staff judge advocate general's office. On November 6, 1999, LTC Griffin mentioned she had been the trial counsel (prosecutor) at Fort Benjamin Harrison in 1988–89 when nine DSs in two training companies were recommended for courts-martial for sexual misconduct with female trainees.

None were actually court-martialed because all but one were discharged under Chapter 10 (discharge in lieu of trial by court-martial). She also said the installation commanding general had granted the trainees immunity from prosecution for engaging in consensual sex so predatory DSs could be identified. That rationale was the same as ours. We had not coordinated with anyone who had worked the cases at Fort Benjamin Harrison, because there had been no data available to us to steer us in that direction.

LTC Griffin later recounted her experiences working the cases:

> Our position was that in a situation between the trainee and drill sergeant, the weight of power is always going to be behind the drill sergeant and you can't root out the abuse

unless you are willing to let the party that is supposed to be protected from the power off the hook. . . .

Initially, all the drill sergeants denied their involvement in senior/subordinate relationships. I was not with the SJA [staff judge advocate] Office when the first drill sergeant was prosecuted. They picked the drill sergeant suspect that they had the best evidence against, gave immunity to the advanced individual training [AIT] student females involved, and after he was convicted, the commander offered to change his conviction to a Chapter 10 if he told investigators what he knew. That was when all the information regarding all the other [9] drill sergeants was confirmed.

This had been going on for several years. Each drill sergeant would pick one AIT student out of their class and she became his "special student." This student would receive extra privileges, get out of certain duty, and ultimately become sexually involved with the drill sergeant. The classes lasted for three months, and after one class ended, the student would PCS [transfer under permanent change of station orders] to her first permanent assignment and the next class would come in and it would begin again.

A very good plan, as the assignments [at Fort Benjamin Harrison] were short-lived, the drill sergeant would insist on silence regarding their relationship while the class was in session, and then the student would leave the area with little likelihood of return. I had been working on the Forward Military Support Element to the Pan American Games on Fort Benjamin Harrison, and came over after the first case. . . .

I became Chief, Military Justice, and had eight cases of drill sergeants on the docket. Clocks were running, and we needed corroborating evidence for the information that the first drill sergeant had provided. At the time there was a discussion as to whether immunity was appropriate for all the female trainees, as there had been some rather inappropriate behavior on their parts. It was my position that this type of behavior was more inappropriate for the drill sergeant than the female trainee, and if the command wanted it stopped they would have to make an example of the leader not the trainee. So we did issue immunity for all the female AIT students.

The fact that these students had all PCS'd far and wide was a factor very much in the favor of the drill sergeants. Once we began the court-martial process, (because while they all took Chapter 10s eventually, these were all General Courts-Martial and we completed many Article 32 hearings prior to the Chapter 10 being offered), we had to fly these women in from all over the world. Our expenses for courts-martial that year busted our budget by tenfold but the command supported it. This was another factor that spoke in favor of the Chapter 10. What the command wanted was the offending drill sergeant out of the Army in disgrace. To continue the courts-martial, we would have had to bring all these witnesses back another time from all over the world. The logistics in these cases was incredible.

The other factor that I found absolutely amazing was that when speaking with these women initially, they were all still very fond of their drill sergeant and were very reticent about talking to me. They didn't want to get their drill sergeant in trouble. We flew these women back for the hearings and before I spoke to them again I had my NCO show them [as a group] the courtroom and give them an idea of where the judge sat and where they would sit as a witness, etc. Then he would leave them in the room for about 30 minutes. By the time I called them in to talk to me, they weren't so happy with their former drill sergeant; they had just figured out that he had been sleeping with all of them. They thought they were the only one.

We were successful in discharging every one of these drill sergeants from the Army with the exception of one. This was a case in which we could only locate one trainee. I referred to it as the "always have more than one trainee versus one lying drill sergeant" rule.

My favorite case was the one where the female student said she had some pictures, would they be helpful. I said, "Yes, please bring them when you come to see me." She showed up with a picture of the drill sergeant and her at her parents' house (she had taken her boyfriend home to meet the folks!) and the other picture was of the drill sergeant naked as the day he was born lying on a couch with a glass of champagne in one hand and a cigar in another. When the defense counsel came in to plead his case (the drill sergeant had hired a civilian attorney, who was actually the head of the local reserve JAG office), I listened politely, and then slid that picture of his client on the couch face down across the table. He picked it up, said nothing, slid it back, and we had a Chapter 10 request in two days.

LTC Griffin's observation of the relationship between the trainee and the DS was insightful. She said:

There's another point that supports the issuance of immunity to the female trainees. When I spoke with these women, some of them had sexual relations a few times with these DSs. Some of them had sexual relations numerous times. What each of these women called these men was drill sergeant not by their first name, not Bob or Mike or Sam, but drill sergeant. Even after intimate sexual relations and being separated for months or up to over a year, every one of them called them drill sergeant. I think it's easy to see where the power in this relationship lies.

This was another indication to me that APG was not an aberration. The type of activity we had uncovered had been going on for years in the Army, was widespread, and was common knowledge. And this Army-wide problem was never solved, let alone addressed by the Army senior leadership, to the best of my knowledge.

I drove down to Martin State Airport in Baltimore on June 3 to pick up LTG Miller and two officers who accompanied him. Back at our headquarters, we spent about four hours with LTG Miller. He started by telling us this inquiry was his primary duty and involved looking into dereliction of duty regarding sexual harassment, but did not include misconduct or felonies. He said his inquiry would look at policies, command climate, and actions taken at all initial entry training installations in the Training and Doctrine Command. He was to have access to all information unless limited by the lawyers.

I thought again for the umpteenth time, "Great—this is just what we need. Another investigation not looking at sexual misconduct."

I then presented the same briefing I had given all other visitors. I concluded the briefing by providing him a list of people I thought should receive some form of punishment, and I provided a recommended punishment for each. Before he departed, I gave him 12 large file folders full of information with additional details.

After LTG Miller departed, I called my contacts in the offices of Senator Mikulski, Senator Sarbanes, and Representative Ehrlich to let them know an announcement would be made the next day on my reassignment to FORSCOM.

My reassignment was officially announced on June 4, and it even appeared on CNN. The Army also announced MG Guest's impending retirement. MG Guest sent me an email, which read, "Just saw you on TV in California about your assignment—they sure planted a seed since two-star assignments aren't announced on CNN—anyway congratulations." BG Tom Dickinson was announced as my replacement as chief of Ord.

I put out a press release that included, in part:

> Being the Chief of Ordnance has been a professionally rewarding experience for me. We accomplished much as a corps in the past two years to support our Army as it enters the 21st Century.
>
> I have a deep appreciation for the many talented and dedicated military and civilian personnel we have in the Ordnance Corps family. They prove over and over again that they lead the way in getting the job done to the line, on the line and on time by accepting and mastering every challenge.
>
> While there are some recent concerns about a small fraction of our corps, the reputation for hard work and diligence established by the Ordnance Corps for the past 185 years is intact. The majority of our Ordnance family, from the drill sergeants teaching and training our newest soldiers, to those who support and maintain our Army in the field, to those who are designing and testing the Ordnance Corps and the Army for the next millennium, are honest, dedicated professionals.
>
> As I move on to my next assignment, I thank you for a rewarding tour as your chief and I will continue to wear proudly our regimental crest anywhere in the world.

One of the individuals we had recommended be disciplined submitted retirement paperwork to leave the Army on August 1. I called LTG Miller and told him I could not stop the paperwork and that it was his call. He told me he would see the Office of the Department of the Army Inspector General (DAIG) report the next week. He concluded the call by saying we had done "good work."

It was also noted in media reports that USAF GEN Ralston allegedly had an affair with a CIA employee while he was married to, but separated from, his wife. This was very similar to what MG Longhouser was accused of doing. However, the secretary of defense came out in support of GEN Ralston's nomination to be the next CJCS.

MG George Friel, the new general court-martial convening authority, turned down requests for discharge in lieu of court-martial from DSs Kelley, Moffett, and Gunter. I was getting tired of dealing with these cases and would have preferred to have had the discharges approved. With a Chapter 10, these three would get essentially the same punishment as from a court-martial, with the exception of jail time. Most importantly, they would be out of our Army.

I went to Fort Lee, Virginia, on June 10 to attend meetings and the Quartermaster Center and School change of command event at which MG Jim "Chickenman" Wright would take command from MG Tom Glisson on Sergeant Seay Field.

I talked with GEN Hartzog at the event. He told me he thought LTG Miller was the right person to do the investigation of the chain of command but that he still had not received a letter from GEN Ronald H. Griffith about what would happen to me.

Secretary West sent a memorandum to GEN Griffith and Sara Lister on June 11 directing the disestablishment of the hotline on June 13. The reason given was that it had served its purpose and "there are indications that the hotline may have been used by some callers for purposes not consistent with its original mission." The letter went on to say that "effective immediately, no anonymous allegations . . . will be accepted over the hotline. . . ." The secretary also announced that within 30 days a new phone number would be established to provide assistance to victims of sexual harassment and sexual abuse.

The media provided an update on the trials in Germany and at Fort Leonard Wood, Missouri. At Fort Leonard Wood, SSG Todd J. Belle was found not guilty of sodomy and adultery charges but found guilty of other counts. He was demoted one rank to sergeant and fined $2,000.

He was the seventh NCO at Fort Leonard Wood to be found guilty by a court-martial panel or plead guilty to sexual misconduct with female trainees. At the Darmstadt Training Center in Germany, SFC Fuller, the second NCO court-martialed from Darmstadt, was found guilty of rape and forcible sodomy. He was sentenced to five years in prison, reduction to Private E-1, forfeiture of all pay and allowances, and a dishonorable discharge. SFC Fuller's lawyer, an Army major, was quoted as saying, "Anybody who says that race is not an issue in this trial is wrong."

On June 13, I flew down to the Pentagon for a meeting with MG Siegfried to get his feedback on his visit to USAOC&S. The session was like listening to a horror story. MG Siegfried had absolutely nothing good to say about my command. I sat there, took notes, and didn't say a word—it was so shocking. He said he had told GEN Hartzog APG was so bad, he needed to visit us.

The bottom line to me was that MAJ Gibson had correctly assessed the situation—I was being set up to be the fall guy for the Army. MG Siegfried told me Secretary West, GEN Reimer, and Lister had asked him "to give them a head."

When I was at FORSCOM, a fellow Ord GO stopped by my office on April 22, 1998, and commented that he knew sexual misconduct was going on at APG before I got there. It appeared to him that once the scandal became a major public spectacle, the Army didn't want to sacrifice a combat or combat support officer (from infantry, armor, artillery, etc.) and instead sacrificed a logistician (from Ordnance, transportation, quartermaster, etc.) like me. This was not the first time I had heard that comment. I don't know for sure whether it was true, but it was further enhanced when MG Joe Bolt, the commander at Fort Jackson, was made a three-star.

After meeting with MG Siegfried, I attended a meeting on a proposal to reduce the rank structure in explosive ordnance disposal teams. Major (MAJ) Donna Alberto was the Ord personnel staff officer in the Pentagon working our personnel policy issues. She always did outstanding work. In her email following this meeting, she said, "I just want to thank all of you, especially the very busy MG Shadley, for attending the various councils over the last three months. I cannot tell you how critical your presence was in achieving the final recommendation for the Ordnance recommendation that will go forward to the Deputy Chief of Staff for Personnel and the VCSA. Those proponent general officers who did not attend . . . sent a message to the [Army leadership] that did not play out in their favor during the wrap up session."

I had been and would be criticized for not spending more time at APG, but this was a good example that demonstrated that if you were not at the table when decisions were being made, you could not influence the outcome.

When I got back to APG in the afternoon, I assembled our senior officer and NCO leadership team. I related what MG Siegfried had told me. They had shell-shocked expressions on their faces. Even if we didn't believe everything MG Siegfried had said, we had an obligation to look at each of his points and assure ourselves we were not overlooking anything. We later concluded we weren't.

A US Army news release announced Command Sergeant Majors (CSMs) Jerry T. Alley and James C. McKinney would serve as co-acting sergeants major of the Army pending the disposition of the case of James McKinney's twin brother, former Sergeant Major of the Army (SMA) Gene C. McKinney.

I would get to know CSM Alley when Ellie and I moved to Fort McPherson and were neighbors with him and his wife, Debbie. CSM Alley and I often talked about the APG sex scandal and NCO misconduct. At one of our neighborhood socials, he told me that when he was a DS at Fort Polk, Louisiana, "70 percent of the NCOs were dirty." He was in the first group of nonvolunteer DSs. The statement CSM Alley made that really struck home was, "The good NCOs cleaned up Fort Polk."

Debbie Alley was in our kitchen with Ellie one hot summer day when I put ice cubes in Remington's water dish. She said, "I always thought you were a good guy, now I know it." She was a dog lover.

On another occasion at Fort McPherson, COL Steve Garrett told me the same things we had uncovered at APG had been happening in 1991–93 at Fort Lee, Virginia.

In retrospect, I believe June 13, 1997, as I left my meeting with Siegfried, was when I said to myself that being in the Army had just ceased to be fun.

17

★ ★

KERMIT THE FROG SAYS:
"IT'S NOT EASY BEING GREEN"

Aт a meeting with Representative Ehrlich in his Washington, DC, office on June 16, I presented him with the Ordnance (Ord) Corps Samuel Sharpe Award for his support of the Ord Corps and Aberdeen Proving Ground (APG). In our conversation, he mentioned that during his visit to APG on December 11, 1996, he had been impressed with our first sergeants but the drill sergeants (DSs) as a whole were not impressive. He based his impression of the DSs on their appearance and conduct during the session. These were the DSs who were at APG when the scandal went public.

Lieutenant General (LTG) John E. Miller called me to discuss his proposed disciplinary action for the commander of the battalion (Bn) in which Staff Sergeant (SSG) Delmar G. Simpson had served. I told LTG Miller I thought it was too late to do a relief (fire the commander) because the regularly scheduled change of command was just a few days away.

David Dishneau's *Associated Press* article from June 20 reported on motions relating to the case of SSG Herman Gunter, and it contained something I did not know. "The [defense] attorneys laid the groundwork at a hearing Thursday by playing a 15-minute videotape of a December [1996] videoconference involving Maj. Gen. Michael J. Nardotti, Jr., the Army's chief prosecutor. On the tape, Nardotti urged subordinate prosecutors to share with Army inspectors the details of pending sexual misconduct cases, including defendants' names. On the same tape, prosecutors were told that in cases where it was unclear whether the sexual acts were consensual or nonconsensual, they should consider them to be nonconsensual."

This was enlightening because we were taking the heat for charging DSs with rape when it was the Army's official position and exactly what prosecutors were supposed to do. Also, if prosecutors were giving alleged perpetrators' names to the Office of the Department of the Army Inspector General (DAIG), then that office should have been able to track common threads and determine the origins and current methods of operation of noncommissioned officer (NCO) players in the Game.

We published news release #41 announcing Sergeant First Class Tony E. Cross had been referred to a general court-martial (GCM). The release reported:

> Cross is pending trial on charges of violating a lawful general regulation by wrongfully socializing with trainees (four specifications), failing to report prohibited relationships (six specifications), sodomy (one specification), and adultery (three specifications). These charges involve a total of eight female soldiers and the incidents allegedly happened between February and December 1995.

On Monday morning, June 23, I attended the graduation of a heavy-wheeled vehicle mechanic course at Edgewood Arsenal (EA). The student class leader presented me with a letter signed by all 40 members of the class commenting on the great job DS Scott Carry did in instilling in them the qualities and traits of leadership. The most telling sentence in the letter was, "He wouldn't let us settle for anything but our best efforts." I sent DS Carry a personal note congratulating him on a job well done.

The new DSs in our organization were top-notch.

The Gamble trial concluded late in the afternoon because he had pled guilty to 23 of 38 charges as part of a plea deal whereby he would testify against others. He was sentenced to 10 months in prison, total forfeiture of all pay and allowances, reduction to private (E-1), and a dishonorable discharge.

Jackie Spinner was at the trial and reported the next day about SSG Wayne A. Gamble being in the Game. In a discussion about SSG Gamble's wife, Jackie Spinner reported in the *Washington Post*:

> His wife, Marilyn Gamble, interrupted the proceeding, shouted out in anger after Gamble apologized to other family members but not to her. "He apologized to everyone but his own wife," Marilyn Gamble said. "He didn't apologize to me." She waited in a grassy area outside the courthouse for her husband to be taken to prison. "I want to see him off," said Marilyn Gamble, 35, who lives in Richmond. "That's the last thing I want to see—him in handcuffs."

SSG Gamble had agreed to testify against SSG Vernell Robinson. Allegedly, SSG Robinson was happy because he only got 6 months in jail and the "rat" SSG Gamble got 10 months.

COL Dennis M. Webb gave me an email he had received from Lieutenant Colonel (LTC) Paul Meredith, who would be taking over as the commander of the 143rd Ord Bn at EA. At the Pre-Command Course at Fort Leavenworth, Kansas, LTC Meredith had asked General (GEN) Dennis J. Reimer, who had been a guest speaker, why he thought APG happened. GEN Reimer replied to him in an email:

> In regard to what went wrong at the Aberdeen Proving Ground, I think there are three major issues that we as an institution must address. First, we were slow to pick up on the change of values in society. We failed to realize as quickly

as we should have the change in the value base in society. Consequently, we did not emphasize values as much as we should. Secondly, the reorganization that was completed a couple of years ago did not give us the oversight capabilities needed for this important function. We cut too much out of the training base and therefore did not have the leadership that we needed to do the critical training mission. This, coupled with the reduction in civilian work force and the borrowed military manpower requirements, gave us an organization that was flawed in its basic design.

I mentioned the fact that we cut too many chaplains but that was just one of the issues. I believe that people were well meaning in designing this organization [of CASCOM] but we didn't realize how critical it was and we cut too much. Lastly, we had a chain of command that went bad. Most of the problems we associated with one company and involved the company commander. When we have a basic leadership failure like this, it is hard for us to get the information that we need to know that it is going on.

I thought GEN Reimer was spot-on with points one and two, but point three was oversimplified—it was more than just one company and one installation.

On June 25, I called LTG Miller at Fort Leavenworth after he requested we talk about the action he was proposing to take against the commander of the 143rd Ord Bn. He reported that GEN William W. Hartzog had discussed the matter with the secretary of the Army (SA) and had decided to suspend the commander to allow him time to respond to a proposed General Officer Memorandum of Reprimand (GOMOR). To me, this was the height of political BS.

I sent an email to Major General (MG) Ken Guest the next morning and included: "[The Bn commander] is scheduled to change command on 2 Jul 97. We are in effect, possibly, going to extend him to see if we should relieve him. This is going to be almost impossible to explain. At best we could take him out of command four days early."

That same day, I had received word that LTG Miller had gotten his charter from GEN Hartzog to investigate the chains of command throughout the Training and Doctrine Command (TRADOC).

This was confirmed in a letter I saw dated the next day from GEN Ronald H. Griffith to GEN Hartzog. It gave GEN Hartzog the authority to take action on commanders in the grade of colonel (COL) and below and on NCOs down to company level. It did not specify whether this applied only to US Army Ordnance Center and School (USAOC&S) or all TRADOC schools

GEN Griffith went on to say in the letter that he had to report any proposed actions to Secretary Togo G. West before they could be taken. I thought to myself that surely the SA has more important things to do than worry about what kind of action was being taken against a captain or an SSG.

I spent about an hour and a half in an off-the-record discussion with Jackie Spinner, who had spent the past two days with two training companies. She was working on a *Washington Post* story about SSG Simpson. Ed Starnes sat in on the meeting to keep me out of trouble. She said she'd also like to talk with me later that coming fall. She thought part of the problem was that Army recruiting was off track—the Army was making it sound as if the young men and women were joining a college.

Jackie Spinner and I would occasionally be in touch over the years, and I had the utmost respect for her. She would help other reporters get their stories straight during the courts-martial, and when I offered to give her one of my general officer (GO) unit coins, which cost about $2 each, she refused because she was not allowed to accept gifts.

There was an article by Dana Priest in the *Washington Post* titled, "Army Panel Destroyed Data on Sex Survey." It appeared to support my suspicion that some Siegfried panel members were manipulating the information to get the results they wanted. The article contained:

A secretive Army panel looking into sexual misconduct destroyed some highly sensitive data it had collected from a survey of 9,000 troops, and a researcher who worked with the group has accused it of "gross fraud, waste and abuse during the course of the investigation. The panel's apparent intention is to suppress this information in order to avoid making the Army look bad . . . ," said Leora Rosen, who works in the department of military psychiatry at Walter Reed Army Medical Center.

According to Rosen and other military sources, the questions were reviewed by all panel members, including its leader Maj. Gen. Richard S. Siegfried, and Sara Lister, assistant secretary of the Army for manpower and reserve affairs. Rosen and other sources said Siegfried and other Army officials had spoken at a panel meeting about the possibility of destroying all the survey data, not just the most controversial questions, in order to control how the results were interpreted

GEN Hartzog called and told me he had decided to issue a GOMOR to the 143rd Ord Bn commander for not knowing about sexual misconduct. He directed me to delay the change of command scheduled for July 2, suspend the commander, and put in the executive officer as the acting commander. GEN Hartzog also directed I provide the commander legal assistance to help him with his response to the GOMOR.

I immediately contacted COL Buzz France to arrange for legal counsel, and I called in LTC Meredith, the incoming Bn commander, to let him know what was going on. LTC Meredith's family was already in the area, getting ready for the ceremony next Wednesday.

After I received the fax, COL Webb, Chaplain Smith, and I drove down to EA to inform the Bn commander of the decision. I called GEN Hartzog on the way back up to APG. I told him I had just executed the dumbest order I had ever received in 29 years on active duty in the Army.

We were delaying a change of command so someone in the Pentagon could decide if the commander should be relieved almost 10 months after everyone in a leadership position above him knew about the incident. GEN Hartzog said he had told GEN Reimer twice that day, "This really sucks pond water."

Earlier that day, Secretary of Defense Cohen had announced that Senator Nancy Kassebaum Baker would head his advisory committee on gender-integrated training and related issues in the military. In his statement, Cohen said, "The problems at Aberdeen and elsewhere have raised questions about the success of gender-integrated training and about the treatment of women in the military."

At our weekly Crisis Action Team meeting, we spent a lot of time going over the status of all the remaining legal actions. I regretted that Brigadier General Thomas R. Dickinson would not have a clean slate when he took command. I had hoped he could have come right in and started the healing process after all the surgery was done, but it was just not to be.

I called MG Guest to let him know my end-of-tour award ceremony would be a private ceremony before the change of command on July 10. The media would be at the change of command event, and it would likely create a stir if the commander who allowed the sex scandal got an award.

We produced my last update, number 27, to the vector report on July 8. The crux of this report was that we provided 41 lessons learned that we felt the Army should consider. This report contained 17 single-spaced pages giving the daily timeline of all major events since the spring of 1996. A detailed accounting of the actions taken included 670 interviews conducted by local CID and 1,413 requests for assistance answered by other CID resident agencies. It documented our 10 primary and 14 supporting actions to identify the systemic causes of the scandal and our 16 actions to prevent any further trainee abuse.

I also covered in detail the status of each case. We had 19 legal actions completed: 1 officer, 8 DSs, 9 instructors, and 1 unknown. Of these 19, there were 12 guilty and 7 not-guilty verdicts. We also had 5 courts-martial of DSs announced and were working with another 2 minor cases still in the investigation phase.

The final vector report I saw after I left APG was dated November 24, 1997, and the only three changes from update 27 were in the disposition of cases chart: (1) All 5 DSs in the courts-martial process when I left were found guilty. One received non-judicial punishment under the provisions of Article 15 of the Uniform Code of Military Justice, 2 were found guilty by GCMs, and 2 were discharged in lieu of trial by courts-martial. (2) The 2 minor cases under investigation were handled by non-judicial punishment, and the subjects were found guilty. (3) The final accounting of the 26 cases (both felony and non-felony) processed through the legal system at APG were 19 guilty and 7 not guilty.

GEN Hartzog called and said he would not sign the Legion of Merit end-of-tour award (a lesser award than the Distinguished Service Medal) for me because "the investigation was not over." He said he didn't want to contribute to any possible embarrassment to the Army. Since no one had ever told me I was being investigated, I found all of this interesting. He said he would swing by Forces Command (FORSCOM) and present it to me after things calmed down. That never happened.

This really didn't bother me. Like most leaders, I never did anything just to receive an award. I figured if you did the right thing and worked hard, those things would come naturally. Worrying about getting awards or promotions was counterproductive.

The change of command ceremony started at 0900 on July 10, and everything was perfect—the weather was great, the troops looked magnificent, the Army Materiel Command Band added their usual superb musical talents to the ceremony, and the marching of the soldiers was by the book.

The outgoing commander usually stays behind after a change of command to say farewell to attendees who want one last good-bye, while the incoming commander goes to a different location to greet the people he or she will be working with. The last person through our receiving line was Paul Valentine from the *Washington Post*, who wanted a statement for the story he would write on the ceremony. To add a little humor to my brief remarks during the ceremony, I quoted Kermit the Frog to sum up the last few months: "It's not easy being green." Valentine included this quote in his story, and he did start the article with the most important part of my talk:

> [MG Robert Shadley] said today the Army post endured "tough times" in the glare of national publicity about widespread sexual misconduct. But he said the acts of a few should not besmirch the entire institution. "Through it all, we've always tried to do the right thing . . . I know it hasn't been easy."
> . . . Shadley told a reporter, "Ninety-nine and forty-four one-hundredths percent of the people at Aberdeen do great work every day. They're good, honest, God-fearing people who have done no wrong. We often lose sight of all that good that is done by all of these people."

I was able to have a short discussion with retired GEN Jimmy D. Ross, who honored us with his presence at the ceremony. I had to break off my conversation with GEN Ross to sit in on a meeting GEN Hartzog was having with the suspended commander of the 143rd Ord Bn. After the meeting, GEN Hartzog and I had a private session. He concluded by saying I had done a great job and that I should go to FORSCOM and lay low for a year.

Ellie and I stayed in guest quarters through Friday night to watch the moving truck being loaded. We pulled out of the main gate at APG at 0830 on Saturday morning, July 12.

We arrived at the quarters of MG Jim and Carol Wright at Fort Lee, Virginia, and Carol had lunch ready for us. Our Shih Tzu, Remington, got along with their two Scotties, Maggy and Stoney, as well as with Hurricane, a mixed breed Jim had rescued during Hurricane Andrew. Jim and Carol were our best friends in the Army, and it was great to start to unwind. Jim was the quartermaster general, so we spent a long time talking about what I thought he should do to prevent a scandal.

We left Fort Lee on Sunday morning and arrived at Fort McPherson, Georgia, in late afternoon, where Major Dave Bullard, my new executive officer, was waiting with keys to our guest quarters. I was looking forward to the new job, the new location, and putting APG behind us—but only the first two would occur.

18

★ ★

I SEE THE BUS COMING

THE JULY 21, 1997, EDITION OF *U.S. News and World Report* contained a short article saying the Siegfried report and the Office of the Department of the Army Inspector General (DAIG) report would be delayed. The last paragraph contained: "Some members of Congress and other Army watchers are waiting to see if the reports, by the Army inspector general and a specially convened Army panel, hold appropriate leaders responsible and recommend corrective action. If not, Congress could do so itself." The accountability agenda was becoming the preeminent force upon the Army senior leadership.

I spent my first week at Forces Command (FORSCOM) getting in-processed, meeting the members of the organization, and getting our quarters set up. We occupied one-half of a large, brick duplex that was on the National Historic Register.

I did call and email what was left of the old team at Aberdeen Proving Ground (APG) to keep current on the legal actions. I was interested in seeing how all this would play out. The FORSCOM staff provided me the daily *Early Bird*. Paul Boyce, the Criminal Investigation Command (CID) public affairs officer whose help had been absolutely outstanding for many months, kept me posted on other issues.

On July 17, the president nominated General (GEN) Hugh Shelton to be the next chairman of the Joint Chiefs of Staff. The nomination of GEN Shelton made GEN Ralston another part of the collateral damage of the hotline.

Secretary of Defense (SECDEF) William S. Cohen was reported to have said in an article by Yumi Wilson in the *San Francisco Chronicle* on July 22 that "he has not ruled out the controversial idea of segregating recruits by gender during training to reduce sexual misconduct." Again, the APG sex scandal was highlighted along with Tailhook. It was apparent that our problems would be recognized for several years as one of the military's landmark scandals. The agenda about the role of women in the military had taken its place as a close second to the accountability agenda.

Paul Richter in the July 24 *Los Angeles Times* talked about no high-ranking officers being punished at APG. Secretary Togo G. West was quoted as telling reporters, "The question was . . . whether the chain of command or some leaders either didn't act when they should have or failed to know information that they should have known." Richter wrote, "But the panel reached a conclusion that many have held for some time: the officers far removed can't be held responsible for what went on in two battalions at Aberdeen and that singling others out would not help remedy the broader problem. Major General Robert T. [sic] Shadley, the former commanding general . . . is unlikely to be tarnished."

An article in the *State* newspaper of Columbia, South Carolina, reported that GEN William W. Hartzog said there would be tighter selection criteria for drill sergeants, possibly a week added to basic training, and likely new values and human relations training. GEN Hartzog was reported to be in favor of keeping gender-integrated training. Reflecting on the APG scandal, he predicted "the Army would be stronger for it." He also said he was convinced that the level and type of sexual misconduct at the Maryland base was not present at other training bases. I remained amazed that senior Army leaders, at least in public, failed to acknowledge that sexual misconduct was not isolated to APG.

On July 29, the Army announced that the panel formed by Secretary West and chaired by MG Steve Siegfried and the DAIG would release their reports in September. Speculation in the media was that the Army hoped to minimize the damage to its image by releasing them at the same time.

It now appeared that protecting the image of the Army was the overriding, number one agenda for the Army senior leadership.

Colonel (COL) Tom Leavitt, who headed the DAIG investigation, called me on July 30 and said he had 500 copies of the DAIG report locked up, waiting for Secretary West to approve the release. I asked him for a copy, but it would be months before I received one.

COL Leavitt also asked me two questions: (1) "Who made the decision on the hotline?" I told him all I knew was that on November 6, 1996, at a meeting in Lieutenant General (LTG) John A. Dubia's office, Sara Lister directed me to set up the hotline. I said that was the only guidance or direction the Department of the Army gave me. (2) "Who decided all former female trainees should be interviewed by CID?" I responded that CID made the recommendation to our Crisis Action Team. We all thought it was the right thing to do, and I asked CID to do it. I reminded COL Leavitt that our first objective was to take care of the victims, and to do that, we needed to find out who and where they were.

Secretary West's memorandum of July 30 directed "the Assistant Secretary of the Army [for] Manpower and Reserve Affairs—Mrs. Lister—to oversee the review and implementation of the recommendations of the panel's report and the Inspector General's special investigation. Both the report and special investigation are scheduled for release in September." Secretary West continued, "Prior to that release, I want to take the findings and recommendations and finish a plan for implementation." He gave Sara Lister until August 20 to get the plan finished.

This made sense. If Secretary West and GEN Dennis J. Reimer

would be testifying before Congress after the documents were published, they would have to report on what the Army was doing to solve the problems.

In a page-one article in the July 31 *New York Times*, Eric Schmitt reported of the two forthcoming Army reports, "Neither inquiry, however, held anyone accountable for the problems, said officials familiar with the reports." This would later turn out to be true because neither addressed the accountability issue.

Although I was no longer at APG, media attention continued to follow me. On July 31, *Washington Post* reporter Dana Priest called my office because she supposedly knew what GEN Hartzog had decided to do in my case. I immediately called GEN Hartzog. He was on leave, but his aide told me GEN Hartzog had sent recommendations on three or four individuals to GEN Ronald H. Griffith and Secretary West, but nothing was said about me. BG Gil Meyer was also on leave. His deputy, COL Robert E. Gaylord, recommended I not talk with Dana.

I probably should have talked with her to find out what she knew and where she had gotten her information, but I was still being a loyal subordinate and doing what I thought was best for the Army.

"Commander at Aberdeen Exonerated" was the title of Dana's article on page one of the *Washington Post*. I thought, "So much for 'go to FORSCOM and lay low for a year.'"

This article, as well as similar articles in other media, was bound to send the accountability agenda folks into a tizzy. The Priest story included:

> Four officers in the two battalions would be disciplined and the disciplinary action was described by an official yesterday as "nothing earth shattering" and may not affect the future careers of all the men involved. . . .
>
> No one in the chain of command was found culpable of any serious charge, such as dereliction of duty. . . . The deci-

sion, which was made by Army Secretary Togo D. West, Jr. yesterday, was not to be made public until next week. . . .

Gen. William Hartzog, who is in charge of the Army's Training and Doctrine Command and who decided on the disciplinary measures after a seven-month review, exonerated Maj. Gen. Robert D. Shadley . . . in part because he believed Shadley had no inkling of sexual misconduct on the base, an official close to the investigation said. When Shadley did find out, he launched an investigation immediately. . . . "He [Shadley] did what he was supposed to do," the official said. "You would have expected the stuff to bubble up, but it didn't. None of the bells rang."

A *New York Times* article countered that no decision had been made about my punishment. The *Washington Post* and *Army Times* would report the same thing. The *New York Times* article stated, "Army officials, speaking on the condition that they may not be named, said that General Hartzog had no authority to exonerate General Shadley—or recommend punishment for him—and that the matter was still under investigation at more senior levels of the Pentagon." This was correct per the letter I saw from GEN Griffith to GEN Hartzog dated June 26.

I would later learn this was all just a trial balloon being floated out to see how Washington would react if the Army did nothing to me.

Later in the day on Saturday, I received Paul Boyce's weekend fax of news articles, and it included an interview with GEN Reimer that would appear in the August 11 edition of the *Army Times*. Two exchanges stood out.

Q: Some criticism of Aberdeen wasn't only who knew what but who should have known what. What are you telling leaders at the unit level about what they should know? In other words, how do you address the issue of accountability?

[GEN Reimer:] The issue of accountability is one that is

very important to us. I really believe that there is a certain level in the chain of command that should have known about this issue. . . . I think we have shown that we are going to hold people accountable for their actions. . . .

Q: What do you say to critics who say those higher in the chain of command should be held accountable?"

[GEN Reimer:] I would say to them: look at the facts, what we have done. I think you have to look at the total situation. We have done a very thorough analysis of that, and I believe we have fixed accountability at the right levels.

COL Gaylord called me later in the day and said the Army leadership was "livid about the accountability issue" and that Secretary West was going to write letters to the editors of the major newspapers.

On Thursday, August 7, Secretary West's letter was published in the *Washington Post*:

In response to Richard Cohen's Aug. 5 op-ed column, "No Bells at Aberdeen," it must be reemphasized that no decision, tentative or final, has been made regarding accountability for sexual misconduct at the Ordnance Center and School at Aberdeen Proving Ground. I have not been briefed on the Training and Doctrine Command's recommendations, and this fact was made clear in unambiguous terms to one of The Post's reporters after the erroneous article "Commander at Aberdeen Exonerated" appeared on Aug. 1. Apparently, without checking the facts, Richard Cohen used this article as the basis for his column.

The fact is that the Army—the first to disclose the situation at Aberdeen and to institute procedures to understand and correct the situation—has proceeded cautiously in this matter with due regard for the harm that could result to all concerned from hastily drawn conclusions.

We would hope that others, in their reporting and commentary, would exercise equal caution.

The trial balloon had lost its air very quickly.

Later in the day, COL Dennis M. Webb, commander of the 61st Ord Brigade, sent me an email to let me know GEN Hartzog had given his approval for the change of command for the 143rd Ord Bn and that it had happened in a small ceremony that very morning.

I was at Fort Riley, Kansas, to visit FORSCOM units and then attend a retirement dinner for LTG Bill Carter. Before the dinner, GEN Hartzog said he wanted to talk to me in private. The conversation would be the first of many in a very confusing, very emotional roller-coaster ride. I didn't take notes during our conversation, but I did write down everything as soon as he left. We talked about a lot of things, including what was going on currently with the folks at US Army Ordnance Center and School and whom he was recommending disciplinary action against.

My final note to summarize my take on this conversation was, "West is going to throw me under the bus."

Later in the evening, COL (Retired) Fred Hepler told me one of the active duty generals at the dinner who was very well connected had said, "the politicos are out to get Shadley." I had lost track of the number of times I had heard that warning. It had to be true.

When I got back to Fort McPherson, I decided to call GEN Hartzog on Saturday to see if I had correctly interpreted what he had said. We talked for almost an hour, and when I hung up, I was more confused than before I had dialed his number. I asked him point blank, "Am I to be the fall guy?" He said I wasn't and that Secretary West said I was okay.

He said, "Cohen, however, is an unknown" and "some people still want your hide." This supported the contention that the "politicos" were behind this.

It was apparent to me SECDEF Cohen was weighing in on the decision process. I based this conclusion on the signal that Secretary West thought I was okay, yet there was nothing the uniformed four-stars could do regarding SECDEF Cohen's decision. While I found it hard to reconcile in my mind, it did appear that Secretary West and GEN Reimer were being told to do something they did not necessarily agree with.

GEN Hartzog went on to tell me his plans were to keep this from being a big deal so I could eventually be promoted and serve as a three-star.

I hung up shaking my head. In the span of less than 48 hours, it looked as if I were going to be either forced to retire or promoted—or something in between.

I called Lieutenant Colonel (LTC) Gabe Riesco on Monday morning to see how he was doing at APG. LTC Riesco said a major had told him former Captain (CPT) Derrick A. Robertson had had problems with sexual misconduct in Korea, but the Bn commander didn't do anything about it. Without even asking for such information, I had been told three of the convicted or alleged perpetrators—CPT Robertson, SSG Simpson, and SFC Cross—were allegedly known to have had sexual misconduct in their backgrounds but no one had done anything about it.

I left FORSCOM headquarters at noon to travel to Fort Lee, Virginia. I was to attend the Logistics Triad meeting the next day, following MG Ken Guest's retirement ceremony. I had a chance meeting with Senator Paul Sarbanes while waiting for my flight at the airport in Atlanta. He said I had done a good job at APG and then offered, "Bet you're glad to be out of there." I told him I only wished I had been able to close out all the legal actions before I left.

After MG Guest's retirement ceremony, GEN Hartzog came up to me and led me off to the side, away from the crowd.

He said, "Good news—General Griffith told me that Secretary West, General Reimer, and he had agreed they would not ask you to retire."

I told GEN Hartzog I would stand up with Secretary West and GEN Reimer at a press conference and take all the blame if he thought that would help the Army. He said, "Nonsense. You've done no wrong."

I still felt that after the dust settled, the Army would aggressively attack the sexual assault problem by ensuring women were protected and by getting sexual predators out of our Army. The motto of the 1st Infantry Division is, "No mission too difficult. No sacrifice too great. Duty first." I felt it was duty to make the sacrifice if needed.

On Monday, August 18, I talked with Paul Boyce about the weekend news articles and the recent legal actions at APG. He said he suspected there would be two DAIG reports—one public and one private. I would eventually see the public report, but I was not provided a copy of the private one, if there were such a thing, even after I submitted an official request for it.

Major Susan Gibson and I talked about my situation and possible options. As always, she had it nailed. She felt the Army would have to do something to me, and whatever they did would be unfair, but I couldn't do much about it without making a mess.

In a call with COL John A. Smith, I learned the two public reports pending release had been compartmentalized and that Secretary West was controlling the release of information. The release date for both reports was now scheduled for September 11.

At the neighborhood evening social, one of my neighbors, who had experience with DAIG investigations, said they did faulty investigations. He also said it had not gone unnoticed in the field Army that GEN Reimer did not support me.

A very interesting bit of information surfaced in the final day of testimony in the McKinney Article 32 hearing. In an article in the *Washington Post* on August 24, it was reported that at the hearing two days earlier, a witness testified (not about McKinney) that "one of her supervisors in Somalia made sexual advances to her." When asked to be more specific, she said, "it started out with him talking about playing the Game."

So, here we now had Playing the Game mentioned in another courtroom—the third I was aware of, not counting Fort Benjamin Harrison in 1988–89.

I talked with MG Guest on Sunday about a conversation he had had with GEN Hartzog before his retirement ceremony. The conversation proved very useful in helping me understand what was going on. I came to the conclusion that evening that I had been the perfect person for the Army to have had in command to uncover what was going on at APG. I was loyal, and I trusted that the Army would protect me if I did the right thing. In other words, I was naïve.

19

★ ★

THE VIEW FROM UNDER THE BUS
IS UGLY

I CALLED LIEUTENANT GENERAL (LTG) JOHN G. COBURN, the Deputy Chief of Staff for Logistics at the Department of the Army (DA), on Monday, August 25, 1997. We talked business, and then we began talking about my situation. He said he had spoken with General (GEN) Ronald H. Griffith a couple weeks earlier. GEN Griffith had said they were going to give me "a mild letter" for my file. LTG Coburn said he told GEN Griffith if the Army gave me a letter of reprimand (LOR), they would have to give my predecessor, Major General (MG) Jim Monroe, one also. My note on this conversation with LTG Coburn read, "Believe this was Griffith getting input from the 'senior logistician' on the Army Staff."

On August 27, MG Joe Bolt called and said he was now a special assistant to GEN William W. Hartzog for initial entry training. He had been tasked to follow up on now retired LTG John E. Miller's assessment of the US Army Ordnance Center and School (USAOC&S) chain of command. MG Bolt said he himself had come up with some more names he thought should be punished. I told MG Bolt to look at the two large binders of our analysis of the chain of command at USAOC&S we had given LTG Miller.

Colonel (COL) Johnnie L. Allen sent me a detailed report on what MG Bolt had said during his visit to Aberdeen Proving Ground (APG) and on the people he was looking at for disciplinary action. It was now almost 14 months after we first started working the sexual misconduct cases, and the Army was still trying to figure out whom to punish. LTG Miller and MG Bolt would both eventually identify basically the same people for discipline as we had identified months before.

The media reports for Friday, August 29, included another court-room discussion of the Game. Sergeant First Class (SFC) Gary F. Alford, a drill sergeant (DS) at Fort Leonard Wood, was found guilty of several charges and sentenced to two years in prison. SFC Alford was in the Game and was the eighth DS at Fort Leonard Wood to have pled or been found guilty of sexual misconduct with trainees.

I sent Major (MAJ) Susan S. Gibson an email to let her know COL Jim Hatten was the new staff judge advocate at Forced Command (FORSCOM). He had told me he was surprised no one in the Army had asked me to talk about our experience at APG to gather lessons learned.

A senior level (four-star general) conference was held on September 3–4. I was later informed that GEN Dennis J. Reimer announced "a general officer would be reprimanded" over the APG situation.

Monday, September 8, was the beginning of a very chaotic week, though it started off calmly enough.

I found an *Associated Press* article on the CNN website about SFC Parrish, a DS at Fort Leonard Wood, receiving a Chapter 10 discharge in lieu of a court-martial. The article went on to discuss the number of other cases. With a little simple math, I computed that compared to our 10 DSs at APG, Fort Leonard Wood had 17 DSs court-martialed, discharged in lieu of court-martial, or pending court-martial.

GEN Hartzog called and said the Army leadership had just completed another round of briefings with the Department of the Army inspector general and the Siegfried panel. They had decided on additional disciplinary action for some officers and noncommissioned officers at USAOC&S. He volunteered that I would get a verbal reprimand, an LOR, or a letter not for filing my official records. He said he thought the third option was the one they selected, and I would be given "a letter I could hang on the wall."

The Training and Doctrine Command (TRADOC) news service published a piece about adding a week to basic combat training, restoring executive officers (lieutenants) to help the training company commanders (captains), formalizing sensing sessions to get feedback from students, and adding chaplains back into the training battalions. I was very pleased with this because we had been advocating for the return of executive officers and chaplains since October 1996.

During the morning of September 9, GEN David A. Bramlett, the four-star commanding general of FORSCOM, asked me to come up to his office. He said GEN Griffith had just called and announced he was issuing me a letter of reprimand. I asked for what. He said, "I don't know—guess we'll find out when it arrives." I had known since September 13, 1996, this was going to happen, but I guess I had held out hope that it would not.

GEN Hartzog called and said GEN Griffith was personally writing the letter. Any fault on my part was by omission, not commission, which I took as good news because it meant I was not being accused of anything criminal. Secretary Togo G. West was scheduled to brief Secretary of Defense (SECDEF) William S. Cohen, and there would be a big press conference on September 11. GEN Hartzog reiterated his previous comment that Secretary West, GEN Reimer, and GEN Griffith all wanted me to stay in the Army and retire as a two-star. This indicated to me the LOR was purely political so Secretary West could report to SECDEF Cohen that he had disciplined a general and the same message would be sent to Congress.

I talked with COL Hatten about getting legal assistance to help respond to GEN Griffith's letter. COL Hatten informed me Lieutenant Colonel (LTC) Bill Kilgallin would be my defense attorney. I did not know LTC Kilgallin, but we would soon be spending a lot of time together. I called around to check him out, and I heard nothing but positive comments. I felt good that I would have professional legal help. I also wanted to get MAJ Gibson involved in helping me. MG Walter B. Huffman, the new Army judge advocate general, would eventually approve MAJ Gibson to assist with my defense.

I spent the rest of the day and well into the night on the phone and email with friends and associates who thought I was being made a scapegoat. One of the last emails I got before turning in for the night was from MAJ Gibson:

> I read your message. VERY unfortunate, but I can't say I'm terribly shocked. I have been quite disappointed with the Army Leadership's handling of all of the command responsibility issues . . . Hang in there. Those of us who know what happened know that you were the guy who had the courage to stand up and do the right thing.

On September 10, LTG George A. Fisher, my immediate supervisor and next door neighbor, brought the LOR GEN Griffith had faxed to GEN Bramlett to my office. Some of the wording made me very upset. I was committed to fighting the letter just as a matter of principle because GEN Griffith had said I didn't care for soldiers. Those were fighting words. If he had not included that part, I would have wadded it up and thrown it into trash where it belonged. Here's what my LOR said:

1. You are hereby reprimanded for failing in your command responsibility to exercise properly oversight of the training units in the U. S. Army Ordnance Center and School. I am aware of the tremendous responsibilities you had as a General Officer in command. In a time of dwindling resources, you were required to exercise both proponency responsibilities, which required extensive travel to diverse locations, and

command responsibilities. These requirements forced you to set priorities on a daily basis, but you failed to give your most important responsibility—caring for soldiers—its proper priority.

2. Your failure was an act of omission, not commission. Specifically, you failed to conduct an accurate assessment of the command climate within the AIT Training Brigade at Aberdeen when you assumed command. Knowing the severe personnel reductions, you failed to assess adequately the effect of the reductions and provide adequate guidance to your subordinates. You focused on branch and proponency issues at the expense of soldier welfare. As you know, in our Army, command carries with it enormous inherent authority and responsibility. The most important of these is discipline, care, welfare, and safety of soldiers; it is the essence of command. Your failure to make this a top priority matter demonstrates questionable judgment and leadership.

3. I acknowledge your excellent performance in addressing the problem once it was identified. You implemented corrective measures effectively and expeditiously. These measures would have been unnecessary, however, had you been more aggressive early in the exercise of your command responsibilities.

4. This reprimand is an administrative action and is not punishment under the Uniform Code of Military justice. I intend to file this letter permanently in your official military personnel file. I will, however, consider any matters you submit before I make my final filing decision. Acknowledge this memorandum within fourteen days, and absent an extension, submit any matters you deem appropriate within that time.

I began working on a press release because I was sure someone in the Pentagon would leak my name and the press would want my reaction. I sent out drafts to several friends and various public affairs officers I had worked with over the past several months to get their thoughts.

That afternoon, COL (Doctor) Cecily David called to tell me CNN Headline News showed my picture with the story that I had received an LOR.

Representative Bob Ehrlich's chief of staff called to say I was being used as a scapegoat and that the congressman would be calling me.

In the next 15 minutes, I received calls from Jackie Spinner (the *Washington Post*), Tom Bowman (the *Baltimore Sun*), Kris Plant and Corey Assembado (CNN), and Karen Palmer (CBS). I did not speak to any members of the media because I could see no good for me coming out of such conversations.

I called GEN Reimer, and his executive officer said he was behind closed doors. When GEN Reimer called back, he starting talking immediately. The conversation went like this:

> GEN Reimer: Bob, I just want you to know that it was not us [the Army] that released your name to the press.
>
> MG Shadley: Sir. Thanks, but that's not what I called about.

> GEN Reimer: Okay. You did call me. Go ahead.
>
> MG Shadley: I have a press release but I won't put it out until after you and Secretary West have your press conference. I don't think you'll like everything in it, but it shouldn't make you mad. What you and the Secretary have to say is more important than what I have to say, and I don't want to jeopardize what you all are doing.

> GEN Reimer: Bob. That's a class act on your part. I really appreciate it.

A friend in the Pentagon told me later that someone by the name of Brian in the office of Ken Bacon, the assistant secretary of defense for public affairs, was allegedly the one who leaked my name to an individual at CBS News, possibly John Martin. This seemed logical to me, recalling Sara Lister's concern on November 6, 1996, that "some SOB up there [OSD] will give information to a friend in the press to curry favor." As I saw it, this was a violation of the Federal Privacy Act of 1974, but it would take time and money to prove. I never considered legal recourse because this was not about personal gain or being vindictive.

In March 1998, Bacon's office would release personal information about Linda Tripp, a key figure in the Monica Lewinsky–President Clinton sex scandal. As a result of this violation of the law, Tripp received $595,000, retroactive salary and the right to reapply for employment.

My name and picture were plastered all over TV and the internet. Someone had obviously gotten access to the LOR because some reporters quoted from it. The story on the CNN website provided a preview of what Secretary West and GEN Reimer would be stressing in their press conference: "Sources say the studies found widespread, but not systemic, problems of sexual abuse at Army training bases." This theme was also stressed in the official Army public affairs office (PAO) talking points and questions-and-answers (Q&As) email to all general officers (GOs): "Sexual abuse and misconduct are not endemic throughout the Army." That APG was an aberration continued to be the message the Army leadership wanted to impart in order to, as it appeared to me and others, protect the Army as an institution.

Bob Ehrlich called me at home that night and said, "This really sucks." He went on to say I had his full support and that he would send me a copy of the press release and letter to SECDEF Cohen he would be putting out the next day. He asked for a copy of the LOR, and I faxed it to him. Some folks in the Pentagon became concerned about my showing people the LOR because it was obviously a softball and only done to appease the accountability agenda.

Later that evening, COL John A. Smith called and said the Army

PAO received a question from the press: "18 cases at Fort Leonard Wood versus 13 at Aberdeen. What is the Army doing to the commander at Fort Leonard Wood?" I had the same question not only about Fort Leonard Wood but also about Fort Jackson, where even more cases were investigated. If something had been done to those commanders, it certainly was not publicized.

On the morning of September 11, COL Smith reported the Army PAO was getting more questions on why it was just me getting punished. The morning's *Early Bird* contained media articles from September 10 with three stories about the reprimand.

Tom Bowman reported in the *Baltimore Sun* on the APG reprimands, including mine: "Army officials believed that 'they did not do enough' to prevent the misconduct. . . . One Army officer said that reprimands are usually 'career-enders.' One Army officer said he thought Shadley's punishment was too harsh noting that the general tried to correct the problem when it came to his attention. 'Here's a guy who blew the whistle on himself and at the end of the day he gets burned,' the officer said."

An article by Steven Kamarow in the *USA Today* included: "The Army is 'hanging Shadley out to dry' for its widespread neglect, says Susan Barnes, a lawyer and advocate for military women."

Susan Barnes was also quoted in a *San Diego Daily Transcript* article by Chris DiEdoardo as saying, "They came to the conclusion that Aberdeen Proving Ground was an aberration, but based on my conversations with military women I have serious problems with that conclusion. If we are going to take down Maj. Gen. Robert Shadley (the commander of Aberdeen, who received an LOR), whose only offense was to have gone public and done the right thing, let's take down Army Secretary Togo West, Chief-of-Staff Gen. GEN Reimer and have a clean sweep."

Dana Priest's article in the September 11 *Washington Post* indicated she had gotten advance word on the press conference happening later in the day for the release of the reports from the Siegfried panel and Office of the Department of the Army Inspector General (DAIG). In regard to my situation, she wrote: "The decision to reprimand Shadley was controversial within the Army. While many women in Congress have insisted that the Army hold its leaders accountable for Aberdeen, many high ranking officials in the Army . . . had previously recommended he be exonerated. 'Somebody wanted a head,' one Army official at the Pentagon lamented yesterday, referring to the reprimand."

An interview with GEN Reimer and the *USA Today*'s editorial board appeared in the September 12 edition. It emphasized Susan Barnes's concerns about how forthright senior leaders were being about the sexual misconduct issue. The first Q&A was:

> Q: Based on the findings of your Aberdeen report and the survey, are their other military facilities with similar problems?

> GEN Reimer: We don't have any indication there is another installation where we have the problems of the magnitude of Aberdeen. We've had problems at other bases; they are well known. We've been very forthright on that. But if there was a problem, it's been fixed.

But as the number of cases that continued over the next several years would show, the problem had not been fixed at these installations.

On September 11, the much-awaited press conference was held in the Pentagon to herald the release of "The Secretary of the Army's Senior Review Panel Report on Sexual Harassment," dated July 1997, and the "Department of the Army Inspector General Special Inspection of Initial Entry Training Equal Opportunity/Sexual Harassment Policies and Procedures December 1996—April 1997" report, dated July 22, 1997.

Secretary West and GEN Reimer went first, and they were followed by MG Steve Siegfried and LTG Jerry Bates. I watched the press conference on the television in my office with members of my staff. We then listened to Q&As via audio feed from the Pentagon. I didn't see or hear any of the presentations by MG Siegfried and LTG Bates.

Secretary West and GEN Reimer did a very nice job of laying out what the two reports said and the comprehensive Human Relations Action Plan to fix things. Again, I was pleased the Army's report contained essentially all of the main recommendations we had submitted months ago.

The next day, I received the transcript of Secretary West's and GEN Reimer's prepared remarks and the Q&A session, so I was able to compare my handwritten notes to the official Army document.

The Army's position was made clear within the first few minutes of the press conference when Secretary West said, "The panel's members have briefed and have expressed the view that what happened at Aberdeen was an aberration. The sexual abuse—sexual abuse—is not endemic throughout our Army. Sexual harassment, however, continues to be a problem."

GEN Reimer then reinforced Secretary West's comments in his prepared text by saying, "I would start out by saying as the secretary has already alluded to, that there is a difference between sexual abuse and sexual harassment

Secretary West then came back to the lectern and announced the disciplinary actions: "General Griffith, the vice chief of staff, has issued a proposed letter of reprimand to a general officer in the TRADOC chain of command."

In the Q&A, several of the first questions focused on me being made a scapegoat and APG as an aberration:

> Reporter: The opposite of the scapegoat question. The IG report in particular cites several failures of oversight and other problems at TRADOC. Could someone explain the rationale why accountability at TRADOC didn't go higher than perhaps the CG [General Hartzog]?

> Secretary West: You don't know that because we haven't identified the general officer. [This was technically correct

because the leak of my name probably came from the Office of the Secretary of Defense.]

It was obvious the Pentagon press corps had the same concerns as Susan Barnes: that this was a much bigger problem than just sexual harassment and that criminal misconduct was indeed widespread in the Army.

The third question and response would prove to be very telling as the unraveling continued:

> Reporter: I was wondering if you might comment more generally about whether or not anybody is being made a scapegoat here, or whether there's pressure for you to discipline somebody who may not have been able to prevent or been in a position to know about wrongdoing, and acted appropriately once they found out.

> GEN Reimer: We have not been driven by pressure on finding a certain level of responsibility. . . . One of the guiding principle's we've maintained and held to throughout is let's do right for the Army here. Doing what is right for the Army involves making sure that we do what's right for the SYSTEM . . . [emphasis added].

This was neither the first nor the last time that protecting the system (i.e., the Army as an institution) seemed to be the primary driver of actions at the senior leader level of the Army.

Earlier, I had sent out my press release and embargoed it until 1400 hours. I felt it was important to remain on the moral high ground. My release read:

> As the Commanding General, U. S. Army Center and Schools; Deputy Commanding General, U. S. Army

Combined Arms support Command; and, the Chief of Ordnance, I was responsible for, among other things, the training and doctrine development for the largest branch in the U. S. Army, which includes the training of over 25,000 soldiers per year in technical skills at eleven locations throughout this great nation.

At one of our school locations, Aberdeen Proving Ground, Maryland, [Edgewood Arsenal was technically a sub-post of APG] we uncovered cases of sexual misconduct, primarily by drill sergeants and instructors who abused their positions of authority and trust. Similar cases we also uncovered at several other Army training schools.

We took immediate action to notify the Army leadership of the problem and initiated an exhaustive investigation to determine the extent of the problem. Ultimately, the Ordnance School at Aberdeen was the only Army school to have every female trainee and former trainee who had attended the school in the past two years located and interviewed. We used the results of these over 1,200 interviews to identify potential victims, to care for these victims, to allow the judicial system to work with regard to alleged perpetrators who were identified by the victims, and to identify corrective actions required to preclude recurrence of the problem.

Unfortunately, neither I nor my predecessor uncovered this problem sooner.

I was the commander when the problem came to light and I accept full responsibility for the decisions I made and the actions I took while in command.

I would like to take this opportunity to reiterate something that I have said many times before. The vast majority of the civilian and military personnel at the U. S. Army Ordnance Center and Schools are hardworking, decent, loyal people who continue to work diligently in training and supporting America's sons and daughters to become productive soldiers and citizens.

Bob Ehrlich called and told me he had put out his press release and had sent a letter to SECDEF Cohen. His press release included:

> Ehrlich, in whose district Aberdeen proving ground is located, called the Army's decision "purely political" and that, "while scapegoating may be convenient nowadays, it is never fair—particularly in the case of a man who has devoted his life to the United States Army."

The congressman also cited comments from the *Washington Post* by an unnamed Pentagon source who said my reprimand was based upon the fact that "someone wanted a head." The press release continued:

> Since the APG sex scandal broke, my office has been in constant, sometimes daily communications with General Shadley. I have been impressed by his proactivity, professionalism, and dedication to duty. To me, it seems unconscionable that the man who brought the situation to the world's attention, who worked tirelessly to ensure that events did not reoccur, should become the ultimate fall guy.

At the end of the day, I received a nice email from a friend, Captain A. Reneé Roberts, who worked in the Pentagon. She said, "I've heard from several people (DOD and Army staff folks) today reference the press conference and you. Just wanted to pass on that many see this as an unjust action."

20

★ ★

THE BUS ROLLS ON

M Y DEFENSE ATTORNEY, LIEUTENANT COLONEL (LTC) Bill Kilgallin, came to visit me on Friday afternoon, September 12, 1997. We had not met, but he said he had seen my picture in the *New York Times* on the flight to Atlanta from Washington, DC.

LTC Kilgallin had worked in the office of vice chief of staff of the Army (VCSA) and had reviewed several letters to and from generals. He said my letter of reprimand (LOR) was mild because it did not say, "have lost confidence in you." He offered that my letter read like, "This guy's a team player, but we need to slap him on the wrist." We talked about how I should reply, and he recommended a one-or two-paragraph "I'm sorry for not being a good commander" type of response.

LTC Kilgallin also confirmed there was an unwritten exception stating that general officers couldn't get underlying information about themselves from the Office of the Department of the Army Inspector General (DAIG). That meant with only the LOR in front of me, I had no idea what I was really responding to.

General (GEN) William W. Hartzog called and related to me a conversation he had just had with GEN Ronald H. Griffith. GEN

237

Griffith had made the decision that I should get the LOR. GEN Griffith had also said he would call me this weekend to tell me how to fight the letter. GEN Hartzog concluded by telling me the letter was a surprise because his understanding had been that the Army senior leadership was leaning toward a verbal reprimand.

I sat by the phone all day Saturday and Sunday, waiting for GEN Griffith to call. He never did.

I appreciated very much Colonel (COL) John A. Smith's sending me the long-awaited DAIG report. As I picked up the quarter-inch-thick document COL Smith sent, I suspected there had to be much more to this exhaustive DAIG investigation, which Secretary Togo G. West had personally reviewed multiple times. But this was all I was ever provided.

I was pleased the report appeared to be based substantially on our vector reports and the other analytical data we had provided to the teams who conducted the investigation.

The report also supported my opinion that several drill sergeants (DSs) not involved in the sexual misconduct did know about it:

> In the IET [Initial Entry Training] environment there is a very close working relationship amongst the drill sergeants and cadre NCOs who serve as instructors. It is, therefore, inconceivable that the NCO corps did not know when one of their ranks was abusing his or her authority. Additionally, the NCO corps was not self-policing in that they chose not to know what was going on and failed to act in accordance with prescribed Army regulations.

I recall my first meeting with the DSs in the 143rd Ordnance Battalion after the scandal was in the national media. They were telling me how they felt disrespected. I said to them, "Okay, if you can honestly say you did not know anything about what was going on, raise your hand." Not one hand went up. I then told them that in my mind, they were just as guilty as the DSs facing charges. Two broke down in tears and left the room. I had my answer.

COL John Smith also sent me a copy of the Siegfried report. It was huge, and I just glanced at it. While it appeared to be a great piece of work on sexual harassment and human relations environments, it didn't address the real problem—sexual felony misconduct by players in the Game.

There was a comment about me in an editorial in the September 22 *Army Times*:

> There may be many in the Army who think Shadley has been dealt an unfair hand. Shadley was the one who blew the whistle on the Aberdeen scandal. He was the one who told America on national television, "We expect that the leader in front of soldiers . . . is truly a leader, and not a lecher." He is the one who asked the Army's Training and Doctrine Command to send him an experienced officer to determine if there are systemic flaws at the school that led to the incidents. And he is the one who will carry the message that commanders will be held responsible for what happens under their command.

Also in the *Army Times*, G. E. Willis wrote an article, "Accountability of commanders is debated," and he noted additional comments by civilian attorney Susan Barnes:

> Susan Barnes said she fears a Shadley reprimand would be perceived as unfair within the Army and lead to a backlash against female soldiers. "What you have is the Army hanging all of this on Aberdeen and singling out the commander for what are, in some cases, career-ending reprimands. To suggest that Aberdeen is an aberration is not consistent from what we are hearing from military women."
>
> "At Aberdeen [Major] Gen. Shadley performed pretty well once this was called to his attention," Barnes said. "By all

accounts, he took action and didn't try to sweep it under the rug . . . So I guess [the Army is] talking about disciplining him because he wasn't vigilant enough . . . But he didn't exactly get assistance from his [Training and Doctrine Command] commanders. . . . These are the same people who took four years to produce a sexual harassment video."

I thought Susan nailed it. I called her on Monday morning to thank her for her support, and we chatted about our appearance together with Katie Couric back on November 8, 1996. Susan went on to say people in the Pentagon had told her in November of 1996 that "Shadley's on a banana peel." She concluded by saying, "[GEN] Reimer doesn't get it."

Letters, emails, and calls of support continued to pour in. Some folks, like Debbie Ward, who was a secretary in the front office when I was GEN Jimmy D. Ross's executive officer at the Army Materiel Command, wrote a personal letter to GEN Dennis J. Reimer and sent me a copy. We had gotten her into an upward-mobility program, and she had moved up from being a secretary to an action officer six grades higher. Debbie got married a few years later and had twin daughters, naming one Ellie because she thought so much of my wife. The last paragraph in her letter touched my heart: "General Shadley served as a mentor and example to many young officers, soldiers, and civilians. What are we to strive for if the best—Major General Shadley—is not good enough?"

I called Debbie to thank her, and she said that based on the way the press was treating me, "little old ladies in Peoria will think you are out groping women." Several years later when I was interviewing for a consulting job, the company told me an internet search showed I was reprimanded for being in a sex scandal.

I received a call from a two-star general in charge of a training base. He told me a colonel came up to him and asked: "Sir, does what happened to General Shadley mean that if I uncover a problem, I could get nailed?" The general replied, "Yes."

I discussed the situation with Major (MAJ) Susan S. Gibson. She followed up with, "You know, I was thinking about the Aberdeen 'anomaly' thing. . . The only anomaly is in your determination to root out every bit of the problem. In effect, they are coming after you for being too thorough—quite a message to send to the field."

I then called my lawyer LTC Kilgallin, and we talked about trying to get information from the DAIG. I told him about my conversation on April 30 with a member of the DAIG staff who indicated that Secretary West had allegedly had them redo the report because they had originally found fault higher up and concluded I should not be punished.

LTC Kilgallin came over from Fort Gordon, Georgia, and we spent almost four hours collaborating on my rebuttal letter. He said he was going to Washington on Friday for a meeting with Tom Taylor in the Army General Counsel's office and would stop by the DAIG office in the morning to pick up any documents they might have about my case.

The next day at the Pentagon, people in the DAIG told LTC Kilgallin that COL Tom Leavitt was out of town and no one else knew where the information on me was located. The DAIG later told LTC Kilgallin they had no information on me to give him.

I worked my rebuttal letter with MAJ Gibson and LTC Kilgallin via fax and phone. A lawyer friend in the Pentagon confirmed several of the facts that had come from other sources and built up my confidence that I had correctly figured the situation:

- Pressure for the LOR came from Congress. (I took this to mean it came from Congress to Cohen and then to Secretary West—the politico chain.)
- The feeling in the Pentagon was that members of Congress, except Bob Ehrlich, were satisfied with the Army action.
- Congress probably wouldn't follow the filing debate because they really didn't understand the process.
- An initial report had come from the DAIG saying there was no more I could have done.

LTC Kilgallin, MAJ Gibson, and I worked on my reply on Saturday.

LTC Kilgallin was now committed to the longer reply option, compared to his first recommendation of a one-or two-paragraph response. When he got back from Washington, he called and said, "We've really got to fight this. You are getting screwed."

Paul Boyce sent me the standard Sunday evening fax of articles. The September 29 *Army Times* had an article that quoted Elaine Donnelly, president of the Center for Military Readiness. "After Aberdeen, the appearance had to be given that something is being done. You call a commission; you call a news conference. But the bottom line is: Togo West is responsible. He should be held accountable."

LTC Kilgallin arrived on September 23, and we worked all day on my rebuttal letter to justify not filing the LOR permanently in my official military personnel file. That afternoon we sent my 11-page, single-spaced letter with 25 enclosures by registered mail to the Pentagon. The enclosures included letters from several supporters on my behalf. We had previously received a verbal okay for a one-week extension, but I still wanted to meet the original deadline of September 23—just in case someone would later say we had not received an official extension, and therefore we were nonresponsive.

I felt relieved at the end of the day. I was confident we had produced a great product and that, at the very least, the LOR would not be filed permanently in my records.

Wednesday, October 1, was the day of the much anticipated Army leadership testimony to the Military Personnel Subcommittee of the House National Security Committee. In a press release, Representative Buyer announced his subcommittee would be conducting the hearing. In the second paragraph, he detailed all the cases at APG, and then in the third paragraph, he stated: "In addition to the cases reported at Aberdeen, there have been a number of cases of alleged sexual misconduct involving drill sergeants and cadre members at other Army training centers such as Fort Leonard Wood in Missouri, and Fort Jackson, South Carolina." I found it interesting that Buyer had used "such as," which to me meant he understood that there were even more locations than the three most publicized locations—APG, Leonard Wood, and Jackson.

I read the official transcripts, and several comments in Secretary West's opening statement bothered me: "The breakdown at Aberdeen Proving Ground was the exception, and the Chair and the Vice-Chair of the Review Panel, and the Inspector General, have called it an 'aberration;' Sexual harassment and, more prevalently, gender discrimination, however, continue to plague our Army, and their eradication will be the focus of our efforts in the wake of these reports."

Secretary West also announced that MG Joe Bolt had been nominated to be the deputy commanding general (DCG) of TRADOC as a three-star. I liked Joe, and we were friends. He was a superb officer. But his command at Fort Jackson was one of the places where DSs had learned the Game, and his command had more cases of sexual misconduct than any other Army training installation.

I received a call on October 6 from a friend in the Pentagon, who said, "West and his folks are really pleased with how they handled Congress and the media. They were two for two—press conference and hearing—more high fives by the Army senior leadership." They may have felt good, but I felt sad for our Army. We deserved better. Most importantly, the young women who were victimized deserved better.

It appeared to me the congressional testimony ended the Army's APG sex scandal, as far as the Army and DOD leadership were concerned. It also appeared to me that agendas such as race, gender-integrated training, accountability, and the image of the Army had trumped our agenda of finding and caring for victims, identifying perpetrators from statements by the victims, and taking action to preclude recurrence. The Army and the DOD had, in my opinion, set themselves on a course to failure in protecting women in the military.

On October 6, it was announced that former Sergeant Major of the Army Gene C. McKinney had been referred to a general court-martial. This would keep Army sex scandal stories going for a few more months.

LTC Kilgallin told me my rebuttal to GEN Griffith's letter had raised a lot of questions within the legal community in the Pentagon and appeared to cast doubt on the DAIG investigation.

CPT A. Reneé Roberts sent me an email on October 21 that included ". . . you getting reprimanded is saying a lot about the lack of morality, ethics, and moral courage at high level. So many that I have spoken to feel that the Army really stepped on it when they used you as a scapegoat. Up to that point actions that were being taken by the Army were considered positive and that we were avoiding a Tailhook. When they slapped that letter on you we did some major damage to our institution. You have done great things for the Army and the people in it. We owe you dearly."

Dana Priest's article, "Persistent Army Gender Issues Cited," in the *Washington Post* on October 22 was about a draft report prepared by Judith Youngman, the chair of the Defense Advisory Committee of Women in the Services after an August visit to Fort Jackson, South Carolina. Priest wrote: "The chairwoman of a panel that advises the defense secretary on women's issues has found that some female recruits and drill sergeants at the Army's largest basic training base say sexual harassment, gender discrimination and alleged sexual assaults remain a problem that local Army commanders have not addressed adequately." It stated further that, "Youngman interviewed 157 men and women, some of whom told her of recent sexual relations between male drill instructors and female recruits. . . . "

I went to Fort Bragg, North Carolina, on Monday, October 27, to address logistics issues. I was scheduled for a 15-minute courtesy call with then Lieutenant General (LTG) Jack Keane, the commander of the XVIII Airborne Corps. The meeting lasted almost 45 minutes, and he was very supportive. It made me feel a lot of pride that one of the Army's premier war fighters thought I had done the right thing at APG. He told me, "No general officer in the Army thinks this [reprimand] is right, and it sends the wrong signal."

In a *Washington Post* article on October 29, Paul Valentine wrote about the Game testimony in the Kelley trial back at Aberdeen: "The game had structure, crimes, sex and concealment with players, props and rules." True.

The one-year anniversary of the historic APG sex scandal news conferences was November 7, 1997. Tom Bowman of the *Baltimore Sun* called on Monday, November 3, and said he wanted to interview me. The story would include interviews with victims and DSs and an interview with LTG Joe Bolt. Ed Starnes told me later in the day Jackie Spinner was also doing an anniversary article. I did not talk with any reporters.

LTG Bolt said he was going to do the interview with Bowman. I recommended he not do it or that he should at least talk with the public affairs people first. We talked about the sex scandals, and LTG Bolt said it sounded like I "was still into this stuff." He also said the gender-integrated task force had visited Fort Leonard Wood recently and they "got bad vibes."

The really disheartening news LTG Bolt passed along was that APG had not received any additional personnel, as everyone had agreed should have been done. A few days later, I sent an email to MG Dan Brown, the new Combined Arms Support Command (CASCOM) commander, and asked if CASCOM had received more money and people as promised. He replied, "The only plus-up so far is in blood pressure."

MG Siegfried and Brigadier General (BG) Evelyn P. Foote visited the Forces Command commanders' conference on November 20 and presented a briefing to our commanders and staff on the secretary of the Army's Senior Review Panel on Sexual Harassment. One chart showed that BG Mary Morgan was one of the special consultants to the panel. Mary told me later that one of the reasons she retired was because of the way I had been treated.

BG Foote told me in a private conversation, "Stuff is going on at other installations even today." Not surprising, because it continued to appear to me that the Army was working hard to make the effects of the problem (negative publicity and congressional scrutiny) go away but not working to solve the cause of the problem (sexual predators).

On my return from travel the day before Thanksgiving, I found a certified mail envelope from the Office of the Judge Advocate General. I knew it had to be the reply to my rebuttal letter. Upon opening it, I found a letter signed by GEN Griffith on October 31, the last day he was on active duty. It read:

1.   I have reviewed your response dated 23 September 1997, with enclosures, to the memorandum of reprimand I issued on 9 September 1997. I agree that you faced significant challenges while in command. However, I remain convinced that in balancing your command and proponency responsibilities, you failed to give soldier welfare issues their proper priority.

2.   I am directing the memorandum dated 9 September 97, your response dated 23 September 1997 (without enclosures), and this memorandum be filed in your Official Military Personnel File.

I was so mad that I could not see straight. Not only had the Pentagon lawyers neglected to call LTC Kilgallin as promised, but GEN Griffith had signed it on his last day on active duty and didn't even have the courtesy to call me or accept my offer to come and see him. I couldn't recall having ever done an adverse action on any subordinate without at least looking him or her in the eye when I did it.

After I had calmed down and made a few calls, I sent the following email to the people who had supported me and written letters on my behalf:

Found envelope in my in-box this morning with letter signed by General Griffith on 31 Oct 97 rejecting my 23 Sep 97 appeal of his 9 Sep 97 letter of reprimand and directing that it be filed in my official file (along with my rebuttal less enclosures). Sort of expected it, but thought I made a strong case. Really appreciate all your support. We'll see what happens next. I just wanted you all to hear it from me and not the rumor mill. Thanks again. Happy Thanksgiving.

The people I sent this to were all smart enough to see this was shoddy treatment. For those who had written letters for me, it was a slap in the face that the enclosures weren't even being made part of the record. I received a reply from everyone, and Ed Starnes's reply summed them up very well:

> Just shows you that justice is not always just. If their own report didn't vindicate you, they don't understand their own report. Also believe that they are not going to pay much more than lip service to working the issues "they discovered." Will be a quick blip on the screen in January for [the McKinney] case, but we'll see the same problem again and again and same report all over. I personally believe you did more than anyone could have expected and that this is bad precedent to set if we want commanders to "take care of soldiers."

I went to see GEN David A. Bramlett, the CG at Fort McPherson. He said GEN William W. Crouch (new VCSA replacing GEN Griffith) had told him about the decision denying my rebuttal, but no one else had said anything. He mentioned GEN Hartzog was also upset. GEN Bramlett said he would call GEN Reimer and ask if the Army were trying to send me a message—like "get out." He said, "The Army has not covered itself with glory in this case."

MG James M. Wright summed up the thought of most of my contemporaries: "We have all lost something in this situation. We have lost confidence that right will win the day and that our leadership will stand behind us when times are hard."

GEN Hartzog called me on December 1 and told me I would probably never know the truth about why I got the reprimand and why it was filed permanently in my records.

When I saw GEN Bramlett on December 5, after he had met with GEM Reimer. He told me GEN Reimer had been uncomfortable talking about me, but had said, "The system was not fair to Bob." I translated "the system" to mean the Army as an institution. GEN Bramlett concluded, "Those in the business know where the problem lay."

On December 9, Paul Boyce faxed me a copy of an article in the October *Washingtonian* titled "The Wrong Stuff," which focused on "denied access to juvenile-crime records, military recruiters are finding it harder to keep the 'evil ones' out." The author, Tom Philpot, pointed out that "[t]he US military doesn't know how many men or women slip into the service each year with arrests or convictions hidden from—or in some cases, hidden by—military recruiters. A conservative figure would be in the thousands, based on recent research for the Defense Department conducted by Dr. Eli Flyer, a manpower analyst." The point that stood out to me was that "a third of all American males are arrested (for crimes that would send an adult to jail) before age 18." Another indication that the future held challenges for Army leaders.

The Federal Advisory Committee on Gender-Integrated Training and Related Issues chaired by retired Senator Nancy Kassebaum Baker published its report on December 16. In her *Washington Post* article on that day, Dana Priest summarized what would likely still be much debated: "A civilian panel appointed by the Pentagon has concluded that female and male military recruits should be segregated during much of basic training and live in separate barracks in order to avoid an erosion of discipline and cohesion."

The specific recommendations affected both Army basic training and advanced individual training. The one recommendation for basic training that would receive a great deal of debate in the coming days was: "At gender-integrated training installations, organize same-gender platoons, divisions and flights and continue gender-integrated training above this unit level."

I signed out on leave for December 20–30. I was glad 1997 would soon be in the books and was looking forward to a new year. I used the time off to figure out, with Ellie, what we wanted to do with our lives going forward.

In retrospect, I had now become one of "them." I had my own agenda now—get the LOR out of my official military personnel file.

21

★ ★

A FAMILY DECISION KEEPS ME IN

OVER THE HOLIDAYS, ELLIE AND I AGREED I WOULD STAY IN the Army until May of 2000, when I could retire as a two-star general (major general). I would also work to get the letter of reprimand (LOR) out of my file. Both of these were a matter of principle and not based on financial or any other considerations. As retired General (GEN) Jimmy D. Ross told me, "You have to be able to look yourself in the mirror," and I needed to do those two things to accomplish that mission. I would play by the rules, not bad mouth the Army in the public domain, maintain the moral high ground, and continue to engage in the eradication of sexual assaults.

Monday, January 5, 1998, was the first work day of the new year. Paul Boyce called to let me know the pretrial hearing for former Sergeant Major of the Army (SMA) Gene McKinney trial would begin on January 6. The trial would go on for weeks and would have a peripheral effect on my case in regard to the race and accountability issues.

Major (MAJ) Susan S. Gibson sent me an email about the Army's new policy requiring commanders to do a command climate survey within 90 days of assuming command. She concluded with, "I guess you were reprimanded for being ahead of the Army."

Reading the Army news release, it appeared the climate survey was for company-level commanders. Brigadier General (BG) Clay Melton, the new director of human resources for the Army in the Pentagon, acknowledged in the Army release that "the survey doesn't adequately address a training environment [like the US Army Ordnance Center and School (USAOC&S)]."

I requested additional details from BG Melton, and in a handwritten note on April 13, he said, "Enclosed is the message requiring the survey and a copy of the survey which became a requirement on 1 Mar. Prior to 1 Mar 98, command climate surveys were optional. I am not aware of any survey requirements that were in place in August 1995 [when I took command at USAOC&S]. Surveys were encouraged but I'm not aware of a specific requirement."

I was really into my job at Forces Command (FORSCOM) now and had put Aberdeen Proving Ground (APG) behind me, except to work with MAJ Gibson and Lieutenant Colonel (LTC) Bill Kilgallin on my request to get the reprimand out of my file. I very rarely sought information about my case, but calls and emails kept coming in, and my status would come up in just about every conversation I had with a military person. I did keep track of articles in the media to get a feel if things at APG and other training centers were getting better as a result of what we went through. It appeared they weren't.

In a January article in the *Newport News Daily*, Lieutenant General (LTG) Joe Bolt was quoted as saying, "I don't perceive those as issues across the training centers . . . those have been fixed" when talking about drill sergeant (DS) misconduct with trainees. This was certainly wishful thinking on LTG Bolt's part.

On Monday, January 26, I sent a six-page letter to GEN Dennis J. Reimer. The subject was my request for clarification of decision and for release of information under the Freedom of Information Act. The letter was hand carried to his office by a friend who was visiting the Pentagon.

One paragraph regarding the request for clarification was the following: "It is my understanding that the original DAIG [Office of the Department of the Army Inspector General] report on sexual misconduct concluded that I had done everything reasonable to preclude sexual misconduct at the U. S. Army Ordnance Center and School (USAOC&S) and that the real blame for the proximate cause of Amy-wide sexual misconduct rested somewhere else. I was told that the original report was rejected by the senior Army leadership and subsequently rewritten. Request copy of the original report(s)."

I also included a long list of materials I was requesting under the provisions of the Freedom of Information Act.

I did receive some material, but nothing of substance. More importantly, I never received a reply from GEN Reimer to my request for clarification of why he approved the LOR issued by GEN Ronald H. Griffith.

Colonel (COL) Jim Hatten, the staff judge advocate at FORSCOM, called the Office of the Judge Advocate General at the Department of the Army (DA) on February 10. He reminded them the Army owed me an answer to my letter to GEN Reimer.

A few days later, I received a letter from the chief, Administrative Law Division, Office of the Judge Advocate General. It said that he had received my letter on January 30, that they had 20 days (not counting weekends and holidays) to respond, and that they had to send my request to several agencies.

On May 6, I would finally receive a letter from the same individual, which contained a smattering of information and a long list of other organizations for me to contact. I thought, "Darn, I sure have missed a lot of weekends and holidays between January 30 and May 6."

Bradley Graham reported in the *Washington Post* on January 28 that GEN Joseph Ralston had been nominated for reappointment as the vice chairman of the Joint Chiefs of Staff for a second two-year term as a four-star. His nomination would have to be approved by the Senate. I joked again that it appeared reporting a sex scandal was more dangerous to one's career than participating in one.

The McKinney trial began on February 5. The proceeding would capture significant media attention over the next several weeks because he was charged with 19 counts of sexual harassment and sexual misconduct. Race and how he was treated vis-à-vis senior commissioned officers would be topics of discussion for several weeks.

On March 13, after deliberating for over 20 hours, Sergeant Major (SGM) Gene C. McKinney was found not guilty of 18 of the 19 charges for which he was tried. The lone guilty verdict related to an obstruction charge. This would later receive some debate: How could SGM McKinney be found guilty of obstructing justice on a charge for which he was found not guilty?

This verdict pleased the African-American community. But it really upset women who felt the alleged victims had been raked over the coals during the trial and that the system (the Army as an institution) had protected SGM McKinney. Several people received the impression that if you were high up enough, you could get off. Some in the media also expressed concern that the verdicts would inhibit women in the future from coming forward to report sexual harassment and sexual misconduct.

SGM McKinney was subsequently sentenced to reduction of one grade to master sergeant and a reprimand. But through a quirk in the law, he would eventually receive the retired pay of a sergeant major.

It would eventually hit me that the McKinney case would inhibit some victims from reporting sexual misconduct and that my case would inhibit some commanders from taking action against alleged perpetrators,

Ed Starnes sent me an email that said, "I'm beginning to believe more and more that you were right about the lack of concern [about the diminished resources at USAOC&S] once the media spotlight has shifted."

I replied, "It's regrettable that the School will never be fixed and, in fact, I predict more schools in TRADOC will follow the [USAOC&S] model—no way TRADOC can function with the [personnel] cuts they are taking and the reduced budgets. Oh well, unless there is another crisis somewhere, I fear Aberdeen will always be considered an aberration by the Army leadership even though down deep I suspect they know differently."

GEN Hartzog and I talked about what he had done in regard to the disciplinary actions for the officers and noncommissioned officers at USAOC&S. He also made a few comments about my case: (1) I had done well for the Army, and with a couple different people in key places, it would have been less painful; (2) GEN Reimer had no political pull; and (3) I would not be a three-star general (LTG). I told him it was okay that I would not be promoted to LTG. He replied, "No, it's not."

On Monday, March 2, MAJ Gibson and I talked about the timing for my formal submission to the Army Board for the Correction of Military Records (ABCMR) to request removal of the LOR from my records. Timing would continue to be a much debated issue for the next several months. Several flag officers would get into trouble, and my name would be mentioned whenever the question of accountability would surface.

Also over the next few months, gender-integrated training would continue to receive a great deal of media attention. The final result would be only some minor changes to gender-integrated training. I shook my head about one of the changes—to have training professionals in the barracks at all times. In an effort to cut down on trainee-on-trainee sexual misconduct back in the spring of 1996, we had done just that, but all it did was make it easier for the DSs to play the Game.

The Army announced the approved list of Army values, and commanders would give each soldier a card containing these values in an official ceremony—Loyalty, Duty, Respect, Selfless-service, Honor, Integrity, and Personal Courage. I thought these were right on. This action would most likely help new recruits and currently serving soldiers understand the Army is different in some ways from mainstream society. This amounted to the code of conduct for which we had advocated more than a year earlier.

In an article in the March 23 *Baltimore Sun*, Tom Bowman discussed the allegation of racial bias made by the Congressional Black Caucus, the National Association for the Advancement of Colored People (NAACP), and individual alleged perpetrators. He stated, "Neither the NAACP nor the Congressional Black Caucus came up with evidence that Army agents were ignoring allegations of enlisted men or officers of other races." Bowman also noted that Staff Sergeant Robinson was the only one at APG to charge racial bias in a court-martial.

Ellie saw Senator Olympia Snowe on *Face the Nation* on Sunday and got fired up. Ellie said, "If she believes what she is saying about protecting the rights of women, then how could she stand by and let you get reprimanded?" In a letter to Senator Snowe, Ellie reminded her of when they had met at APG in 1997. Ellie pulled no punches in her letter:

> My concern is that you and others in Congress allowed certain individuals in the Department of Defense (DOD) to destroy my husband's future in the Army by making him the scapegoat for an Army-wide (if not DOD-wide) sex scandal. There is no person in a leadership position who has come forward on their own to do more to support and defend the rights of women than him. He is the most honest and fairest person on this planet, and he treats all persons—regardless of gender and/or ethnicity—with respect and dignity.

The MG David R. Hale case made the front page of the *Washington Times* on March 27. Rowan Scarborough reported: "Gen. Dennis Reimer, the Army chief of staff, allowed a two-star general to retire honorably in February while he was still under investigation for suspicion of forcing an officer's wife to be his mistress and maid." MG Hale was a bachelor, and the alleged offenses occurred in Izmir, Turkey. When he retired, MG Hale was the deputy to the inspector general (DTIG) at the DAIG.

In a follow-up to this story, Scarborough reported in March 28 that Secretary of Defense William S. Cohen had directed the DOD inspector general (IG) to report on the status of the two-month investigation of MG Hale. He had also instructed Judith Miller, the Office of the Secretary of Defense General Counsel, to conduct an inquiry.

I had never met MG Hale or knew of him, but his case had an indirect effect regarding my appeal of GEN Griffith's LOR and my retirement. I was beginning to see the wisdom of waiting on my appeal until GEN Reimer retired. MG Hale's situation convinced me I would have little chance of being allowed to retire as an MG as an exception to policy (before completing three years of service in that rank).GEN Reimer did not look good for allowing MG Hale to retire honorably, no matter how the Army tried to spin it. So in my case, he would not want to appear to be "soft on another bad general."

MG Hale later became the highest-ranking officer to be court-martialed since 1952 and pled guilty to 8 charges, as part of a plea deal—7 counts of conduct unbecoming of an officer and 1 count of making a false official statement. His civilian attorney noted that he had not pled guilty to any nonconsensual sex charges. He was later sentenced to paying a $10,000 fine and forfeiting $1,000 per month for one year from his retired pay. There would be a few articles in the media comparing how MG Hale (who was white) was treated vis-à-vis SGM McKinney (who was black).

Several months later, it was announced that MG Hale's retired rank had been reduced from two-stars to one-star by Secretary of the Army Caldera.

On March 30, 1998, COL Tom Leavitt called and said I was being called as a witness in the DAIG's investigation of racial bias in the Criminal Investigation Command (CID) investigations. I hung up and immediately called LTC Bill Kilgallin. I told LTC Kilgallin I would never again talk with any IG without a lawyer present and that I wanted him to be with me when I testified.

I called LTC Howard Olson in the general officer management office on April 2 and asked how I would go about requesting to retire as a two-star without three years in grade. He advised that now was not a good time—an obvious reference to the MG Hale situation. This confirmed my take on the situation, so I put that idea on the back burner and continued to assemble supporting documents for an appeal to the ABCMR.

For more than two hours on April 9, I gave my testimony to the DAIG team investigating CID for alleged racial bias. I said I knew of no bias in any of the CID actions. LTC Kilgallin said I did okay.

Russ Childress called on April 17 and reported that 22 people on the staff and faculty at USAOC&s had been let go because of further budget cuts. Instead of getting additional staff, the school was being downsized. Every indicator was continuing to point to no real changes as a result of our efforts at APG.

Tresa Baldas reported in the May 7 *Chicago Tribune,* "The Navy is investigating allegations that a training officer [a 30-year-old enlisted man] had an improper sexual relationship with at least one of his female recruits [an 18-year-old] during boot camp at the Great Lakes Naval Training Center. . . . The allegation comes eight months after U. S. Secretary of Defense William Cohen visited the Lake County base and praised it for being a 'role model' for male-female integrated training and for being free of sex scandals." Baldas also provided data on sexual misconduct "at the four boot camps that train men and women together" during the period 1995–96. The bases cited were Great Lakes Naval Training Center, Fort Leonard Wood, Fort Jackson, and Lackland Air Force Base. More cases at the Great Lakes Naval Training Center would be documented in the media over the next few months.

While at Fort Riley, Kansas, on May 8–9, I visited logistics units and then attended the Fort Riley Logistics Ball. A young captain came up to me and said, "Sir, there's a large group of us who consider you the inspirational leader of our Army today." I was humbled and honored by that remark and thought to myself maybe I had done some good after all.

22

THE FINISH LINE DOESN'T EXIST

THE *Washington Post* ON JUNE 9 REPORTED THAT "DEFENSE Secretary William S. Cohen said yesterday that he and other defense officials will try to dissuade members of Congress from separating women and men in basic training, a provision that has passed the House and will be debated in the Senate. . . . [H]e ordered the services to ensure the sexes are sufficiently separated in the barracks."

The key here was that men and women would be permitted to live on the same floor in the same building. The reporter, Dana Priest, went on to write: "Gender-integrated training was scrutinized after the sexual harassment and rape cases at Aberdeen Proving Ground in Maryland."

The Congressional Significant Activities Report of June 26 stated that "Members [of the Senate] agreed (56-37) to an amendment by Olympia Snowe (R-ME) that keeps the status quo in training." This in effect meant the Senate was supporting gender-integrated training, while the House was siding with the advocates of gender-segregated training.

The 50th anniversary of President Truman's signing the executive order that marked the beginning of racial integration in the military was on July 26, 1998. There were several race-related articles in the media.

The *Baltimore Sun* article included: "The NAACP also reiterated its concern that in the Aberdeen Proving Ground sex scandal the Army unfairly targeted black officers based on complaints of white female recruits. The Army has denied that race was a factor in its investigation."

It was announced that the Department of the Army (DA) office of public affairs had won a Silver Anvil Award in crisis management from the Public Relations Society of America for "the way the sexual misconduct challenge was handled." The society's website proclaimed, "The Silver Anvil, symbolizing the forging of public opinion, is awarded annually to organizations that have successfully addressed a contemporary issue with exemplary professional skill, creativity, and resourcefulness."

A couple of people in the business sent emails to me saying this would have not happened without the work of the Ordnance Center and School and the Test and Evaluation Command public affairs staffs at Aberdeen in support of the Army. I certainly agreed and would have really liked to have seen our team at Aberdeen Proving Ground (APG) more visibly recognized, especially Ed Starnes. Overall, the public affairs aspects of what we went through were well done. It was a shame the multiple agendas deviated attention from the real problem—prevention of sexual assaults.

MAJ Susan S. Gibson sent me an email on August 11 that included:

> I was called by the DAIG [Office of the Department of the Army Inspector General] again today for some follow-on questions on the investigation by CID [Criminal Investigation Command]. They read me my rights, so the interview didn't go very far. Anyway, now I'm looking for a defense counsel before I talk to them again. They have questions about the "immunity" policy and about CID not reading rights to trainees. If this were a normal case, I'd just answer their questions, but it isn't normal. Of course, if I am the next political football in this case it won't matter what my answers

are anyway. Same goes if they are after Don Hayden. As you well know, they don't need any facts to reprimand someone. Why is it that the folks who worked so hard to correct the problems are the ones who end up on the hot seat?

The month of September marked the approaching one-year anniversary of my letter of reprimand (LOR). I continued to focus most of my effort on my job at Forces Command, and I felt our logistics team was making a difference supporting the units in the field to maintain their readiness. I also continued to work on my appeal. Keeping track of sexual misconduct, the investigation of the CID, the congressional commission looking into gender-integrated training, and the status of actions to correct the problems at APG became a hobby for me. I needed one, because my golf game was terrible.

On September 1, I sent LTC Bill Kilgallin an email to discuss the timing for the submission of my appeal. I asked what he thought of the reply I had received from the Pentagon that no one could find any documentation justifying GEN Ronald H. Griffith's decisions. Not a real surprise, but at least we had it in writing.

GEN William W. Hartzog called me on September 11 and said he was calling others and me to chat because his previously announced retirement ceremony was going to take place in a couple of hours. We talked about a lot of things, and when it came to my case, he said I had done no wrong, that the politics came from Secretary Togo G. West, and that I was right about the timing of submitting my appeal—it would be best to wait until GEN Reimer retired.

Major General (MG) Ken Guest told me he had spoken with GEN Griffith at a ceremony in Washington, DC. MG Guest had told him it was a shame about what had happened to me. GEN Griffith responded that he had to issue my LOR because the DAIG had come back to him and said conditions still existed and that I had not done anything to make things better.

I thought, "Great. How do I respond to yet another reason for the

LOR?" First, GEN Griffith had claimed I had failed to conduct an accurate command climate assessment, but had done a good job after I found problems, which I should have found earlier (September 9, 1997). Second, he had said I hadn't made caring for soldiers a top priority (October 31, 1997). Third, now he was claiming I really hadn't done anything after I had found the problem, despite him originally commended me for doing that very thing (September 9, 1998).

I called COL Tom Leavitt at the DAIG and asked him if he could get me any reports about me. He said he would try. He also concurred that I should wait until GEN Reimer retired to submit the appeal. He said my LOR was "pleasant" and that he had seen a lot worse. I never got any reports.

I ran into Brigadier General (BG) Daniel A. Doherty at a meeting in Washington on October 8. He said the DAIG had found fault in his decision to not have the CID investigators read female trainees their rights. He told me several years later he received an LOR for this decision, and I'm sure this was a major contributing factor in his not being selected for promotion to two-stars.

Chief Warrant Officer 3 Don Hayden told me later he also had been reprimanded for the same thing.

Here were two honorable men, working hard to get to the bottom of felony offenses by drill sergeants (DSs) in gangs, and they were the ones punished. They had focused on treating these young female soldiers as the victims they were and had done exactly what the team at Fort Benjamin Harrison had done in 1988–89 in exactly the same type situation.

In the *Washington Times* on November 1, Ernest Blazar wrote about a study of civilian-military leadership relations in Washington conducted by retired COL Lloyd J. Matthews. The last two paragraphs really struck home to me:

> Among the last of [Matthews's] recommendations is that
> U. S. military's four-star officers must be prepared to sound

off publicly if their civilian masters court danger through excessive interference and micromanagement. For many senior officers, the greatest test of courage falls not on the crimson fields of military battle, but rather within the genteel suites of bureaucratic strife.

In a related article, Paul Craig Roberts wrote in the *Washington Times*:

> The day another star meant more to a general than military pride was the day the military lost its leadership. There is nothing the rank-in-file can do when their putative leaders desert them for career reasons.

Paul Boyce and I had an email exchange on November 2 about sexual misconduct cases at three Training and Doctrine Command installations. I said, "The problem as I saw it was that this was bigger than these installations. Little cards listing values and in-process reviews aren't going to get it." I then said, "As we discussed, we have a problem with the NCO Corps, but I've seen no program to address that issue."

I meant no disrespect to the vast majority of the absolutely superb noncommissioned officers in the Army, but even if only 7 percent of the DSs were rotten, then we still had 8 to 10 bad ones causing problems at each gender-integrated training installation.

I called LTC Kilgallin to see what his contact on the Army Board for Correction of Military Records (ABCMR) thought about my case. His source had told him he didn't think timing was a major issue. LTC Kilgallin also confirmed there were seven or so sexual misconduct cases working at Fort McClellan. He also surprised me by suggesting he go talk with GEN Griffith to see if he would write a letter on my behalf to the ABCMR. I told LTC Kilgallin to go ahead because it wouldn't hurt to try.

About midmorning on November 2, MG Ralph Wooten, the

commanding general at Fort McClellan, called and solicited my advice on how to handle the seven or eight sexual misconduct cases he had working. MG Wooten was a superb officer, and I appreciated and respected his call. He said his cases were just like the ones we had at APG. I told him how we had handled things and that I was available any time to help. Over a period of four-plus years, only a handful of officers in leadership positions in the Department of Defense, including MG Wooten and MG Jim Wright, ever asked for my advice on how to handle sexual misconduct cases.

I called BG Gil Meyer to see how the Army public affairs office (PAO) was handling Fort McClellan, but he was out of the office. I talked with his deputy, COL Robert E. Gaylord, and he asked what I had heard. I told him I had heard folks calling it "Little Aberdeen," and he confirmed he had heard the same thing. COL Gaylord said they were going to use the APG PAO model in the Fort McClellan situation. He said the model had also worked at the Disciplinary Barracks at Fort Leavenworth when they had problems with abuse of power by guards against inmates. I couldn't remember hearing anything about that problem.

COL Gaylord also said they would not publish an overall press release on Fort McClellan but instead issue a press release as each case was referred to court-martial. One press release had already been published. That was the same approach I had advocated for with Sara Lister on November 6, 1996. It appeared to me the Army leadership must have learned its lesson about putting out scintillating, overarching press releases and holding press conferences with high-ranking officials on all the major networks.

On November 13, BG Doherty sent me an email that read, "I attended an AUSA [Association of the US Army] luncheon at which GEN Hartzog was the guest speaker. As he spoke about the Army values, he used you and the manner in which you stepped forward in the wake of APG revelations as an example of personal courage. Reminded me of the tremendous respect and admiration I have for you. I am very proud to be counted as a friend of yours."

I was very touched by this unsolicited comment, and it meant even more coming from such a great officer and gentleman. BG Doherty always strove to do the right thing, and like me, he would pay a price. In my reply, I said, "Dan Doherty also stood up and was counted, so the feeling is mutual."

Representatives from the Government Accountability Office (GAO) came to Fort McPherson on November 17 to check into operational support issues. In a side bar conversation, the team told me the GAO had released a report on gender-integrated training. Later, I learned the GAO report also said there was still more sexual misconduct going on than the Army knew about. Another example of very little, if anything, really changing—except a decrease in media coverage.

I talked with LTC Bill Kilgallin on Wednesday, December 3, and learned quite a bit about what was swirling about in the Pentagon regarding my case. The real bombshell he dropped was that we had to, by Army regulation, submit my appeal to something called the DA Suitability and Assessment Board before going to the ABCMR. I had never heard of this board. LTC Kilgallin said only one other time had a general officer submitted an appeal to this board. The COL who ran the board told LTC Kilgallin it sent appeals to the Army leadership, which did not appear to be a good thing for me. When LTC Kilgallin mentioned my name to the COL, the COL said, "He really got screwed."

LTC Kilgallin came over to Fort McPherson the next day. We worked on his upcoming discussion with GEN Griffith when they would meet on December 10, and we redefined our strategy for submitting the appeal. I recommended several points I thought he should cover with GEN Griffith.

On December 15, I took almost four pages of notes as LTC Kilgallin related the comments GEN Griffith had made during their two-hour meeting on December 10.

GEN Griffith had told LTC Kilgallin the LOR had been "appropriate at the time" based on "the totality of circumstances," but he would support withdrawal of the letter from my records based on the fact that it had already served its useful purpose.

In their conversation, GEN Griffith said LTC Kilgallin did not have all the facts, and there was nothing LTC Kilgallin could say that would convince GEN Griffith otherwise. LTC Kilgallin said GEN Griffith was brutally blunt. GEN Griffith said the DAIG team told him, "Shadley had done nothing to correct the problems."

GEN Griffith concluded the meeting by saying he thought the letter should be removed from my records—upon retirement. LTC Kilgallin told me he had sent GEN Griffith a draft, without the retirement caveat in it.

I was glad GEN Griffith would support removal of the letter, but I was upset by a couple of his comments LTC Kilgallin had relayed to me.

If there were facts we didn't know, then why hadn't the Army provided them to us? In a letter dated August 17, the Office of the Secretary of the Army's freedom of information officer, Sylvia M. Chang, wrote to me: "A physical search of the files by the Records Custodian for the Office of the Vice Chief of Staff of the Army was performed. This included all of the safes, file cabinets and desks for records identified in the subject request [my letter of January 26]. No records responsive to the request were found."

I took "totality of circumstances" to mean external pressures from Congress and the Office of the Secretary of Defense. This confirmed my opinion that the letter was done for political reasons so the Army could say in a most public way, "We disciplined a general." This also appeared to be the "useful purpose" served.

The letter of support from GEN Griffith for removing his LOR from my official personnel file was signed on January 7, 1999, and I sincerely appreciated his support. This document would carry a lot of weight in my appeal package.

His letter basically reiterated the original LOR and included, "I acknowledge his excellent performance in addressing the problem once it was identified and censured him for failing to identify the problem sooner." It then stated, "My intention was to ensure that command shortcomings were, in fact, addressed and that the necessary measures had been taken to preclude similar command problems within the Training Brigade [at USAOC&S]. I have concluded that the intended purpose of the memorandum has been served and recommend it now be removed from MG Shadley's Official Military Personnel File."

Some of the wording in GEN Griffith's letter of support didn't synch with what he told LTC Kilgallin, but I had a letter nonetheless.

Because the appeal process had now become like a hobby, I was anxious to see what twist was coming next. I would not be disappointed.

23

★ ★

ONE SMALL PERSONAL VICTORY IS WON

A CLOSE PERSONAL FRIEND CALLED ON JANUARY 12, 1999, TO let me know he was working with the Suitability Evaluation Board that would receive my appeal. While this person could in no way affect the outcome of the board, he could keep me apprised of the appeal's status and other information that didn't violate any rules and regulation.

An official US Army news release on January 15 announced that the Office of the Department of the Army Inspector General's (DAIG's) "investigation of racial bias by the Army Criminal Investigation Command (CID) . . . found that all allegations of racial bias were without merit." The DAIG looked into the allegations of racial bias in Sergeant Major Gene C. McKinney's case and the cases we took to trial at Aberdeen Proving Ground (APG). I called and emailed Brigadier General Daniel A. Doherty of the CID to congratulate him on the finding and to let him know there was never a doubt in my mind that Chief Warrant Officer 3 Don Hayden and his team had been fair to all concerned.

When considering the allegations the CID team had investigated, I did find two sentences in the DAIG report ironic: "It is important to understand that, contextually, the APG trainer-on-trainee investigations took place in a leadership environment sensitive to taking care of soldiers. This value system may have contributed to viewing female trainees as victims." According to General (GEN) Ronald H. Griffith in his letter of reprimand (LOR) and filing determination letter, I had not been concerned with taking care of soldiers. Now the DAIG said I was too caring.

The Junction City, Kansas, newspaper of January 22, 1999, reported on a talk retired GEN William W. Hartzog gave at a dinner in his honor the night before. Regarding Army values, it read:

> Hartzog gave an example of personal courage when he spoke about the sexual harassment scandal that hit the Army at Aberdeen Proving Ground several years ago. "Gen. Bob Shadley displayed personal courage when he went to the media representing the U. S. Army," Hartzog said. "He did what he had to do. He did what was right to him, at a personal cost to him and his career. He did it because it was right. I am extremely proud of Bob and his personal courage."

The *Washington Times* reported that on February 5, the staff had received a copy of a directive from Secretary William S. Cohen ordering the following:

> . . . the military branches to tighten retirement procedures for generals and admirals in order to ensure that senior officers facing misconduct charges are not allowed to immediately retire honorably. . . . [The Cohen directive] gives retirement approval to the service secretary or undersecretary, and prohibits delegation to any subordinate. The order also mandates lines of communication between service secretaries and those investigating an admiral or general.

This new directive meant the secretary of the Army, and not the chief of staff of the Army, would be acting on my retirement request in a few months.

I called my source at the Suitability Evaluation Board at his home on February 15. He provided much useful information on the process my appeal letter would go through. He also said the Congressional Black Caucus was the group that had pushed for me to be reprimanded. My source also reported that he had not seen any documentation justifying the LOR written by GEN Griffith.

I was at APG for meetings on February 19 and had a long chat with Russ Childress. He said the Department of the Army had told the Training and Doctrine Command (TRADOC) to fix the Ordnance Center and School, but not much had happened. All that had been provided to the school was a promise by TRADOC headquarters to look at things.

While in Washington, I attended GEN Johnnie E. Wilson's retirement ceremony at Fort Myer. After the ceremony, I spoke with GEN Dennis J. Reimer's wife, Mary Jo. I mentioned the emotional toll APG had taken on Ellie and the other spouses. The past three-plus years had been harder on Ellie than me, and I thought Mary Jo should know. She did write Ellie a nice note.

On June 30, now lieutenant colonel (LTC) Susan S. Gibson called from Hawaii and said she was being investigated for not having read trainees their rights at APG. It was going on three years since the scandal surfaced, and people were still being investigated for trying their best to do the right thing. I told her I'd do anything to help. LTC Gibson was subsequently vindicated and went on to retire as a colonel (COL) and then continue her service to our nation as a senior civilian government employee. She would serve as the principal deputy general counsel for the Office of the Director of National Intelligence in Washington, DC.

I told my source in the Pentagon I would hand deliver my appeal to him when I was in Washington on July 28. The document I submitted contained a 17-page cover letter with 32 enclosures. My source called me on August 2 and said the package looked good and he'd start the process rolling.

On August 11, COL Jim Hatten, the staff judge advocate at Forces Command (FORSCOM), talked to me about my appeal. I gave him a copy. He asked if I would use a positive response to try getting promoted. I said, "Absolutely not."

Some good news finally arrived on August 26. Another friend in the Pentagon informed me a panel of three-star generals had met to review my appeal letter. The board unanimously supported the removal of GEN Griffith's LOR from my official military personnel file (OMPF). LTC William F. Kilgallin said the appeal with the favorable board recommendation went to the judge advocate general on August 29. They then sent it on the Office of the General Counsel.

Meanwhile, news of sexual misconduct continued to surface in the news. The *Chicago Tribune* reported on August 29, "An Army drill sergeant was placed on administrative duty and charged with 25 counts of sexual misconduct and other improper conduct with 25 female recruits [at Fort Jackson, South Carolina]." I hoped, for the Army's sake, that this would not be dismissed as another "aberration." These allegations seemed to suggest that maybe even Staff Sergeant Delmar G. Simpson was not an aberration.

This drill sergeant was subsequently found guilty of 19 charges and sentenced to 44 months in prison, reduction to private, forfeiture of all pay and allowances, and a dishonorable discharge.

On September 22, 1999, I called Chief Warrant Officer 5 (CW5) Carl Carnes, the chief of staff's administrative officer in the general officer management office. I told him I was sending some information for my file, and he confirmed the appeal was "at legal for review." He very graciously offered to keep me posted on the appeal's progress, and he predicted the letter would be removed from my file.

Frederick A. Taylor, a leader in the Military Order of the Purple Heart, sent me an email on October 12 mentioning he would be seeing President Clinton in November. He asked if I wanted him to mention anything to the president.

I told him I was letting my appeal work through the process, but I said, "I trust your judgment on what needs to be said."

Fred replied, "I know how the system works, and I also know that a 'fall guy' down the line is at times picked to carry the dirty laundry. DON'T CARRY THE DIRTY LAUNDRY FOR SOMEONE ELSE." The capitalization was Fred's, and it hit me that this was exactly what I had been doing.

On October 13, I sent a draft of my request for retirement on June 1, 2000, to CW5 Carnes. The requested retirement date was one month after my three-year anniversary as a major general—I wanted to be absolutely sure I made the three years. CW5 Carnes confirmed I would meet the three-year-in-grade requirement on May 1, 2000, and therefore, I would retire as a two-star. He concluded his email by saying, "Reference your action [my appeal to remove the LOR from my file]. The ASA(M&RA) [Assistant Secretary of the Army for Manpower and Reserve Affairs, Patrick T. Henry] is reviewing and has not made a recommendation to the SecArmy." This was another change in the process as originally told to LTC Kilgallin and me.

More stories about sexual abuse in the Army hit the national newspapers. Bradley Graham reported in the *Washington Post* on October 23, "The Army's top noncommissioned officer in Europe faces charges of kidnapping, sodomy and other offenses for allegedly assaulting a female subordinate during a trip to the German town of Hanau last April. . . ."

It was absolutely amazing that this stuff was still going on. It would be announced later by the Army news service that this command sergeant major (CSM) had been found guilty of fraternization and was fined $1,852 per month for two months and given an LOR. I expected this finding would create a lot of interest. This CSM was white and was sent to an Article 15 hearing, whereas former Sergeant Major of the Army Gene C. McKinney had been sent to a court-martial instead. Perhaps to reduce media attention, the Army timed this press release perfectly— two days before Christmas.

My source with knowledge of my appeal called later in the day and confirmed that my paperwork was out of the normal chain and that the Army was not following its own procedures. LTC Kilgallin and I exchanged emails on the situation. He had been reassigned to US Central Command headquarters in Tampa, Florida. On November 29, he called and offered that we were being slow rolled. We were also in a box—we could not appeal to the Army Board for Correction of Military Records until after this administrative process had run its course.

I was in the Pentagon on December 1 and ran into LTG John M. Pickler, the new director of the Army staff in the Pentagon. He asked me about my appeal, and I told him what I knew. He said he would check on it.

Ellie and I were back in Ohio for Christmas, and I received a call from retired Major General (MG) Jim Wright. He said, "Bobby, I've got bad news. I've been diagnosed with pancreatic cancer, and it doesn't look good." Ellie and I were heartbroken. For the next few weeks, we stayed in close contact with his wife, Carol, to see how things were going.

On February 4, 2000, Ellie and I flew to Houston to see Jim at M. D. Anderson Hospital. When we got there, he looked terrible. He said, "Bobby, glad you are here. Last week would have been too early. Next week will too late." We got all of his personal affairs in order, and he really suffered on Saturday. My best friend in the Army died at 0400 hours on February 6, 2000.

What I had gone through and was going through paled in comparison to what I had witnessed in my best friend's suffering. His only complaint to me was, "Is this all they can do for the pain?"

I looked at him and said, "Jimbo, I think it is."

When we got back to Fort McPherson, I started working on all the emails that had backed up. On February 6, CW5 Carnes had written, "The Sec Army approved your retirement in the grade of MG. I should have retirement orders to you by Tuesday [February 8]. Still no word on the removal of the [reprimand]. Mr. Henry still has it."

I received the orders to retire as a two-star dated February 8, 2000.

On March 22, LTC Kilgallin said he had talked with Tom Taylor in the Office of the General Counsel. As I interpreted it, Henry and the vice chief of staff were trying to decide if my appeal had followed was the correct procedure. He talked about some other issues, which to me confirmed the slow-roll theory. It was what it was. I told LTC Kilgallin I had an office call scheduled with LTG Pickler later in the day and I would ask him to check on the status of the appeal. I did and he said he would find out what was going on.

LTG Pickler called back on March 26 and said he had talked with Henry and others about my appeal. Henry had told him they were waiting for Secretary of the Army Louis Caldera to take action on my retirement request. I told LTG Pickler I already had orders dated February 8, 2000, authorizing me to retire as a two-star. I faxed him a copy.

P. T. Henry was a Senate Armed Services Committee staffer back in November 1996 and allegedly was one of those whom may have been calling for my head. No surprise I didn't get expedited service on my appeal.

CW5 Carnes sent me an email on March 27 that read, "Reference the special action that you have pending, you must have influenced someone, anyway, it finally moved to VCSA [vice chief of staff of the Army, now GEN John M. Keane] office this morning. You will most likely have an answer by week's end." I'm sure LTG Pickler had made this happen once he saw I had had retirement orders for almost two months.

On April 10, CW5 Carnes sent me an email with good news: "The CSA approved removing the information you requested. If you would like, I can fax a copy and the original by mail." I received the fax of the letter signed on April 7 by GEN Eric K. Shinseki, which read:

1. I have reviewed your request to remove the 9 September 1997 General Officer Memorandum of Reprimand from your Official Military Personnel File (OMPF).

2. After careful consideration, your request is approved. . . .

I sent back an email to CW5 Carnes that asked: "No mention of it going into restricted fiche [the OMPF is on microfiche, and there are two levels of access], so it sounds like it is out all together. Am I right?"

CW5 Carnes responded, "You are correct, it is completely removed from your file. I should be able to get a clean fiche in about a week."

He did.

At least now if anyone ever had a reason to look at my OMPF, the LOR would not be there as an official black mark on my record. I would now focus on closing out all the actions still open at FORSCOM and start the transition with MG Terry Juskowiak, who would be replacing me. I accepted a job with Alliant Techsystems in Minnesota, working for one of my old bosses, retired Admiral Paul David Miller.

We had a very nice retirement dinner on the evening of May 1 and a first-class ceremony on Henekin Parade Field on the following day. The FORSCOM commanding general, GEN Jay Hendrix, was very gracious and allowed retired GEN Jimmy D. Ross to officiate at the ceremony. GEN Hendrix presented me with the Distinguished Service Medal, and Representative Bob Ehrlich sent a flag that had been flown over the Capitol in my honor. Ellie received recognition for all her support over the years. Many friends and family came, and we had a very pleasant reception at the community club following the ceremony.

I had special guests at the ceremony: Dr. Coleman and her first-grade class. I had been reading to her classes for more than two years at Arkwright Elementary School, near Fort McPherson. The students at the school were all black and many from single-parent families. I would buy five or six African-American culture books at Barnes & Noble, read one during each of my visits, and donate them to Dr. Coleman's classroom. The kids were great, and it was a pleasure and honor to work with such a dedicated servant of society as Dr. Coleman.

Ellie, Remington, and I rolled out of gate of Fort McPherson at 0850 hours on May 24, 2000, heading to Minnesota to start career number two. However, the Army's APG sex scandal was not completely behind me.

24

★ ★

THERE IS LIFE AFTER THE ARMY

ELLIE, REMINGTON, AND I ARRIVED IN THE MINNEAPOLIS area on the Friday before Memorial Day weekend in 2000. We found a Marriott that allowed pets a couple of miles away from the Alliant Techsystems (ATK) corporate headquarters in Hopkins, Minnesota. The next morning, I put on my sweat suit, stocking cap, and mittens and got Remington ready for his morning constitutional. As we exited the main entrance to the hotel, the first person I saw was a young woman in shorts and a halter top. I realized then that 50 degrees was considered warm in Minnesota. Coming from the upper 80s in the Atlanta area, it was cold to me. I had a lot to learn.

Working for Admiral Paul David Miller again was a pleasure, and I had a very enjoyable and rewarding seven-plus-year career at ATK. Everyone I associated with at ATK was a true professional who saw their primary mission as providing our military and NASA with the highest-quality products.

Ed Starnes kept me posted with the latest articles and reports in the media until he passed away on February 19, 2005. Ed was a wonderful human being and a first-class public affairs officer, and I will be eternally grateful for his help.

Russ Childress retired on September 30, 2001, and Ellie and I stayed in contact with him and his wife, Joan. Russ continued to keep me posted on happenings at Aberdeen Proving Ground (APG) until the Ordnance Center and School was moved to Fort Lee, Virginia, on September 11, 2009. This completed the action that was first discussed in 1997 about moving Ordnance out of APG.

I was able to keep relatively up to date via the weekly *Army Times*; media reports on the internet; and contacts with active and retired Army officers, noncommissioned officers (NCOs), soldiers, and civilians. I kept track of the problems with sexual harassment and misconduct in the Department of Defense (DOD) in general and the Department of the Army (DA) in particular to see if what we had gone through had made a difference.

There was a short entry in the January 15, 2001, edition of the *U.S. News and World Report* that read: "Four years after the Army sex scandal at Aberdeen, Md, the Army has done little to aid harassment victims. The Defense Advisory Committee on Women in the Services (DACOWITS) reveals that some female soldiers don't even know 'where to turn for assistance.' Worse, it's not exclusively a domestic problem: Female troops working overseas are often sexually harassed by 'host nation' brutes."

Steve Vogel reported in the *Washington Post* on April 14:

> The Marine Corps is conducting a major investigation into allegations of sexual misconduct committed by instructors against students at a military base in Missouri, officials disclosed yesterday. A Marine Corps staff sergeant pleaded guilty to sexual misconduct charges in connection with the investigation at a hearing yesterday at the Quantico Marine Base. Three other instructors were charged yesterday with similar offenses, and the Marine Corps continues to investigate approximately 20 Marines.

Vogel went on to report that the base was Fort Leonard Wood, and that "Investigators have interviewed more than 400 former students and instructors at Marine Corps installations worldwide, as well as students and instructors now assigned to the school." It appeared to me that at least the Marines at Fort Leonard Wood were taking this seriously and interviewing former students, contrary to the approach the Army leadership took at that same installation in 1996–97.

Early in the morning on April 17, I received an email from Major General (MG) Terry Juskowiak, who replaced me at Forces Command. It read: "Steve [Koons, my deputy at FORSCOM, who was still in the position] and I saw you on CNN 'Headliners' this morning. It was a report on Aberdeen. They had your picture. Thought you might want to know and make sure Ellie doesn't see it." I assumed they were comparing the Marine scandal at Fort Leonard Wood to the Army's sex scandal at APG. I was too late getting the message to Ellie, because she called to say I was on CNN.

Vogel reported in the *Washington Post* on April 29, "Three more Marine Corps instructors will be court-martialed on charges related to a sexual harassment at a Missouri training base . . . bringing to seven the number of Marines who have been charged." It appeared that six more instructors faced some sort of punishment. With that many involved, I wondered if anyone had checked whether these guys got the idea from their Army counterparts at Fort Leonard Wood? It appeared to me the GAM was still in vogue.

In an article titled "Drugs, Sex and Recommendations" in the *Washington Post* on July 17, Thomas E. Ricks noted, "Sexual misconduct and abuse continue to result in prison terms for service members, but the cases haven't been getting the media attention they used to receive." I took this to mean the media was tired of reporting on the cases and/or the public was tired of hearing and reading about them. In either case, sexual felonies were still being committed.

On September 11, 2001, I boarded a Northwest flight at Minneapolis–St. Paul International Airport for Washington, DC. I had a meeting booked at 1300 at the Pentagon. Shortly after we began our initial descent into the Washington area, the gentleman sitting next to me said he didn't think we were heading east any longer. Then the captain announced, "Ladies and gentleman, we have been diverted back to Detroit. There's an air traffic control problem on the East Coast."

I called back to my office in Hopkins, Minnesota, to ask my administrative assistant to reserve me a rental car in Detroit because I suspected I would have to drive to Washington, DC. It was then that I learned the World Trade Center Towers had been attacked, along with the Pentagon. The gentleman next to me worked for Honeywell, another Minnesota company. ATK was a spin-off from Honeywell. He called National Car Rental directly. He was smarter than I was. I heard him say he wanted a car in Detroit but not within 50 miles of the airport. Ray Bronson, also with ATK, was on the same flight, along with another Honeywell guy. Because of the companies' connection, we had a natural bond. The four of us decided to travel together, as it was now obvious we'd have to drive back to Minneapolis.

Needless to say, there was no car for me at the airport, so the four of us took a 45-minute cab ride to a northern suburb of Detroit, where we had a rental car waiting at a filling station.

The focus of the military, as well as the focus of the whole nation, was now on preventing future terrorist attacks. And rightly so.

At a dinner for General and Mrs. John G. Coburn in Alexandria, Virginia, in honor of his pending retirement on December 1, I sat with then MG Mitch Stevenson, Russ Childress, and other folks from APG who had driven down for the event.

MG Stevenson told me he went to the headquarters of the US Army Training and Doctrine Command (TRADOC) at Fort Monroe, Virginia, for an orientation before becoming the chief of Ordnance. He said the staff there really "bad-mouthed" me for doing a bad job at APG. That comment drove home the point that while just about everyone I ever talked to about APG was very supportive, there must have been an equal number of people who thought I was a real dirt bag.

In October 2002, I ran into then MG Mike Rochelle at a conference in Washington, DC. MG Rochelle had been the installation commander at Fort Monroe from April 1995 to June 1997. He made several nice comments about how we had handled the scandal at APG. I sent him an email on October 23: "Just wanted to drop you a note to let you know how much it meant to this old soldier for you to stop me in the hallway at AUSA [Association of the US Army] and say such nice things about how we handled our challenge at APG."

MG Rochelle responded:

> I feel very fortunate to have witnessed the events we were discussing from the vantage point that I occupied. Moreover, I feel blessed to have observed such men of character and integrity coping with it in the manner in which you did. As I expressed to you, it is a lesson I will never forget. Notable in this experience were General [William W.] Hartzog and yourself, both of whom I was pleased to see last week. The lesson I took away for all this is that we never know who we are "teaching" nor the lessons that they take away. An awesome responsibility. Thanks for being such a great teacher!

With each passing month and year, I received comments such as those from MG Rochelle, who would go on to serve with distinction as a three-star. These comments answered the question for me: Was it worth going through what we did, to try our best to do the right thing? While I had mixed feelings about whether we had really made a difference, I was sure we did the right thing.

The *Washington Times* carried a story in January 2003 that said, "The Army's top brass has concluded that mixed-sex recruit training is 'not efficient' but nevertheless is a policy worth keeping, according to an internal study. . . ." According to the article, the study determined that although the mixed-sex boot camp leads to a disproportionate number of injuries to women and is not efficient in producing new soldiers, that overall the system was working.

Ed Starnes sent me a draft of a new TRADOC pamphlet, "TRADOC Trainee Abuse Prevention Program," on February 3, 2003. The APG sex scandal was implied to be the reason for the document. The seven-page document had less than a full page on "preventive measures." I found it interesting that almost seven years after the scandal started, 29 lines of text on "preventive measures" was all the guidance the TRADOC staff could come up with to help commanders.

One result of a sex scandal at the Air Force Academy was reported in an article in the *Army Times* on December 8: "At a Nov 19 hearing on the Army budget, Sen. John Warner (R-VA), [Senate Armed Services Committee] chairman, asked [Sen. Saxby Chambliss, R-GA], who heads the personnel subcommittee, to study whether the military fails to prosecute sexual assailants and punishes the alleged victims for coming forward." I found this amazing considering what the APG team—and Brigadier General Daniel A. Doherty and his Criminal Investigation Command team in particular—went through for treating female soldiers as victims, for not reading them their rights so they would help us find the perpetrators and other victims.

In the fall, I became a logistics subject matter expert (LogSME) for the Army's Battle Command Training Program at Fort Leavenworth, Kansas. My boss at ATK and the corporate lawyers gave me clearance to participate in this work. Retired MG Tim McLean was the "lead" LogSME, and he needed a second person to work an exercise in Japan in January 2004. Much of the success of Army logistics units (later called sustainment units) would be a result of MG McLean's marvelous ability to coach, teach, and mentor.

I served my first assignment as a LogSME from January 22 to February 1, 2004, with Lieutenant Colonel Dave Gaffney and his team from the US Army Battle Command Training Program at Camp Zama, Japan. I really enjoyed working with the great logisticians from the US Army Reserve deployed to Japan to participate in the annual US–Japan Yama Sakara exercise. I felt I still had something to contribute to the

Army and looked forward to doing more senior mentor work.

As part of my senior mentoring, I presented professional development classes from February 2004 to September 2010, including during the two trips I made to Iraq. I gave the class to the leadership of each unit with which I worked. Setting and enforcing high ethical and moral standards was a key topic. I used my experience at APG to emphasize that it was up to each and every person to make sure the group as a whole was doing right. I gave this class to hundreds of officers and NCOs over a six-year period. Almost every time at the end of the class, someone would come up to me and say, "Sir, that was great. I wish you had been able to give that talk to my unit before we deployed the last time."

Jane McHugh and Gina Cavallaro reported in the *Army Times* of February 23, 2004:

> Distress calls have poured in since Defense Secretary Donald Rumsfeld on Feb. 6 ordered an investigation of reported sexual assaults in Operation Iraqi Freedom. . . . The inquiry, according to David Chu, undersecretary of defense for personnel and readiness, will focus "on the effectiveness of our policies and programs, the manner in which the department deals with sexual assault and our effectiveness in precluding such assaults in the first place."

This issue would be a topic in the media for years. It would continue to point out that the DOD and the Army were still struggling with the problem we had identified almost eight years prior. In my mind, until the emphasis is placed on solving the root causes of the problem instead of trying to make the problem go away, the military will continue to attack the effect and not the cause of the problem.

As a follow-on, Deborah Funk wrote in the *Army Times* on March 8: "Senior lawmakers have warned Defense Department officials that if they don't deal with the problem of sexual assaults in the ranks, they will do it for them."

An editorial in the March 15 edition of the *Army Times* appeared to support my take on the situation: "There's growing anger on Capitol Hill and in society about reports of American troops sexually assaulting fellow service members in Iraq. Too bad service leaders don't seem to share that anger. Instead, they seem more focused on damage control than on cracking down on sex crimes in the war zone and at home."

An Army news service release of June 3 stated: "The Army is devising a policy that will reemphasize that all offenses of sexual assault must be reported to the Criminal Investigation Command. . . . " A task force, established by Acting Secretary of the Army Les Brownlee, spent three months looking at the Army's policies and programs on sexual assaults. The article went on to report that Secretary Brownlee approved the task force recommendations and they were briefed to the House Armed Services Committee on June 3.

Returning home on August 5, 2004, from a trip to Picatinny Arsenal, New Jersey, where Ray Bronson and I had talked with our tank ammunition customer, I was in line at Newark Airport and noticed retired General (GEN) Dennis J. Reimer in front of me. I tapped him on the shoulder, said hello, introduced him to Ray, and we chatted briefly. It turned out we were all on the same flight back to Minneapolis.

As I got off the plane in Minneapolis, GEN Reimer was waiting for me at the top of the jet-way and asked if he could talk to me about APG. I said of course. He proceeded to tell me that I had done nothing wrong at APG, that neither he nor the Army did anything to help me, and that he wanted to apologize to me. He went on to talk about my personal courage and how the APG team had done the right things. I was really touched and got teary.

A few days later, I sent GEN Reimer an email that read: "Words cannot express how much our conversation at the airport on this past Thursday evening meant to me. I apologize for getting emotional, but you touched me very deeply. . . . Please accept my sincerest appreciation for your most kind words...."

GEN Reimer replied on August 20, "Bob—it was great to see you and most of all to learn that you are doing so well—I appreciated the opportunity to express my feelings because this issue was one that has bothered me for some time—as I said I felt like you were very much part of the solution to the problem we faced and I wasn't able to come thru for you like I should have."

I would continue to keep reading about the ongoing issues of gender-integrated training, women in combat, continuing cases of felony sex offenses at many Army installations in the United States and overseas, and how victims of assault were treated.

But August 5, 2004, marked the end of the Army's Aberdeen sex scandal for me, personally.

EPILOGUE

As I collected information for this book, I looked back at over 16 years of events since that fateful phone call I received in Germany in September 1996. I wonder if it were all worth it. My answer is a simple, yes. Yes, because we did the right thing.

I do regret that the senior leadership in the Army did appear to be more interested in making the problem go away in the media and more interested in the influence of Congress and the organizations with agendas than they were interested in solving the problem of women being sexually assaulted in the military.

After August 5, 2004, I kept abreast of sexual misconduct in the military via open media sources, my work as a senior mentor, and comments from friends still on active duty.

It appeared to me the GAM never stopped.

The headlines of the April 11, 2005, edition of the *Army Times* screamed, "Another Sex Scandal." Jane McHugh, who visited Aberdeen Proving Ground (APG) in 1996, went into detail about the cases of sexual misconduct at Fort Knox involving a company commander and four drill sergeants (DSs). In addition, there was a scandal at Fort Bliss where DSs were selling physical training (PT) "insurance" to trainees to ensure they passed their PT test. These DSs also sold snacks and soft drinks to trainees at exorbitant prices—another example of DSs abusing their position for personal gain.

Major General Mark Hamilton's comments to me in 1998 about recruiter misconduct going unnoticed by the media became ironic when

I opened the September 6, 2006, edition of the *Army Times* and read an article by Martha Mendoza of the *Associated Press*:

> More than 100 young trainees who expressed interest in joining the military in the past year were preyed upon by their recruiters—raped on recruiting office couches, assaulted in government cars, and groped en route to entrance exams. A six-month Associated Press investigation found that more than 80 military recruiters were disciplined last year for sexual assault and misconduct . . . with potential enlistees.

In the December 15, 2008, edition of the *Army Times*, Gina Cavalarro wrote an article whose headlines blared out, "Drill Sgt Sex Scandal—12 Fort Leonard Wood instructors preyed on recruits." The article began:

> Drinking parties. Sex in the laundry room. Social dates and text messaging. Sex in a truck. In a bathroom. And in the barracks. Between February 2007 and November 2008, 12 drill sergeants and advanced individual training instructors at Fort Leonard Wood, Mo admitted in court-martial proceeding to engaging in such forbidden sexual and social relationships with trainees.

There would be no more DS sex scandals at advanced individual training (AIT) sites in the Army after 2008 because DSs were done away with in name. Platoon sergeants were now in charge of trainee soldierization in AIT.

In February 2011, I received a letter from a friend that included an article from the *St. Louis* (Missouri) *Post Dispatch* titled, "Sexual assaults persist at Fort Leonard Wood."

In 2011, I noted with interest the similarities of what we went through at APG and the scandal with the football program at Pennsylvania State University. It appeared to me Jerry Sandusky abused his position of power, and the senior leadership at the university appeared to have been more interested in making the problem go away than in stopping a sexual predator.

The military has not fully identified or corrected the problem of sexual misconduct. The *Army Times* reported in March 2012 that Secretary of Defense Leon Panetta said, "DOD finally will create a uniform, centralized sexual assault reporting system across the services—something mandated by Congress in 2009." This was initially directed, as I understand it, by Congress in 1989.

Sexual assaults continue. The *New York Times* on March 8, 2012, reported: "Defense Secretary Leon Panetta estimated the number of attacks in 2011 by service members on other service members—both women and men—was close to 19,000, more than six times the number of reported attacks." This means over a quarter of a million service members may have been assaulted since this issue was brought to the Department of Defense's (DOD's) attention in 1996 by our team at Aberdeen.

As we were putting the finishing touches on this book, I read with anguish about the scandal at Lackland Air Force in 2012, where female trainees once again were victimized by lechers.

At APG, we treated young women as victims no matter what their role was in any sexual misconduct with the noncommissioned officers and the one captain. At a minimum, they were victims of bad leadership by the DSs and instructors who, in most cases, were twice their age. Even if the trainee initiated the contact, their leaders should have shown them the right way and not participated in the wrong way.

I believe we at APG did accomplish some good things: (1) We facilitated the departure of one captain and several noncommissioned officers who abused their positions of power to sexually assault their subordinates, and in doing so, we made our Army a better place. (2) We offered numerous suggestions to the Department of the Army to help correct the problems we uncovered, and many were accepted and adopted by the

Army as its own. (3) We showed there were some leaders who were willing to stand up and do the right thing, even at their personal expense. We stayed true to our objectives and made finding and caring for victims job number one.

Even though some will disagree, I know down deep we did the right thing.

I was able to use the experience I gained to help leaders in units for which I served as a senior mentor from February 2004 to September 2010. I was, however, disappointed that very few in senior leadership positions in the Army or the DOD had asked for my opinion on how to avoid sexual misconduct and the resultant scandals.

If I were ever asked, here are a few of the points I would make:

I know the women appointed to positions to work the sexual assault problem are working hard. I mean no disrespect with the following comment: As long as the military keeps putting women in charge of the prevention of sexual harassment/sexual assault, these problems will be seen as women's issues and not military issues. Prevention of sexual assault is not a personnel or human relations issue; it's a force protection issue. It needs to be handled in units by the same staffs working to prevent injury and death by improvised explosive device attacks, terrorist attacks on facilities and people, etc. Prevention of sexual assault is a force protection and unit operational readiness issue, plain and simple.

When working with the Department of Defense budget, be careful when cutting people who facilitate a commander's ability to gain situational awareness of what is going on in echelons below him or her. It is essential that all support mechanisms are in place for soldiers and that information to decision makers is not impeded.

Prevention of sexual harassment training needs to be continual and frequent. It appears to me a vast majority of sexual assaults find their beginnings in sexual harassment.

Using sex to get ahead should not be tolerated. Women need to police their ranks just as men must do for theirs.

Anyone found guilty of sexual assault and other felonies should be drummed out of the Army. No second chance, no mercy—just as the Army handled drug users beginning in the 1980s.

In my simplistic mind, the key to the prevention of sexual harassment and sexual felonies in the military is for every soldier and civilian—regardless of gender, ethnicity, religion, or rank—to be a keeper of the standards.

If a soldier or civilian sees someone doing something that even appears to be wrong, he or she needs to call the offender (male or female) out on it. Give that person a chance to stop, unless it is so bad, higher ups need to know right away. If that person doesn't stop, report them to their leadership.

Leaders must do the "tough right" and not the "easy wrong." They must act on concerns brought to their attention, and their subordinates need to know it's okay to take their complaints through other channels to get resolution.

The enforcement of the highest tactical, technical, ethical, and moral standards is up to every soldier and civilian in the military. If we are going to stamp out misconduct of all types, every person must enforce the standards. If you are not part of the solution, you are part of the problem.

Finally, as I have repeatedly said, 99.44 percent of the people in our Army are good people, working hard every day to do the right thing, and we cannot and should not let a few bad actors drag down all the good people.

As we enter another era of reduced federal budgets, I suspect the military will have the same or similar problems. The temptation will be to cut executive officers, equal opportunity staff officers, chaplains, counselors, and so on to save money. This has been proven to be a flawed course of action.

The recent decision to open combat arms military occupational specialties to women will add another dimension to the challenges of leadership, but these challenges are not insurmountable.

In my experience, problems with "things" in the military are only minor. The "people" aspects of the organization are where the major problems come from. If you take care of your people—reward the good and correct the bad—the mission will get done, and the leaders and their organizations will succeed.

PRINCIPALS

ABELL, Charles S., Lieutenant Colonel (Retired), Staffer, Senate Armed Services Committee, Washington, DC

ALBERTO, Donna, Major, Ordnance Personnel Staff Officer, Office of the Deputy Chief of Staff for Personnel, Department of the Army, Washington, DC

ALLEN, Johnnie L., Lieutenant Colonel later Colonel, Deputy Commandant, US Army Ordnance Center and Schools, Aberdeen Proving Ground, Maryland (succeeded LUTTRELL)

ALLEY, Jerry T., Command Sergeant Major, US Army Forces Command, Fort McPherson, Georgia, later Co-Acting Sergeant Major of the Army

BACON, Kenneth N. (Ken), Presidential Appointee, Assistant Secretary of Defense for Public Affairs, Washington, DC

BARNES, Susan G., Civilian Attorney

BATES, Jared L. (Jerry), Lieutenant General, the Inspector General, US Army, Washington, DC (replaced by JORDAN)

BECKWITH, Charles E., Colonel, Inspector General, US Army Training and Doctrine Command, Fort Monroe, Virginia

BLACK, Scott C., Lieutenant Colonel, Legislative Counsel, later Chief, Investigations and Legislative Division, Office of Legislative Liaison, Office of the Secretary of the Army, Washington, DC

BOLT, William J. (Joe), Major General later Lieutenant General, Commanding General, US Army Training Center and Fort Jackson, South Carolina (succeeded SIEGFRIED), later Deputy Commanding General for Initial Entry Training, US Army Training and Doctrine Command, Fort Monroe, Virginia

BOYCE, Paul, Jr., Department of the Army Civilian, Public Affairs Officer, US Army Criminal Investigation Command, Fort Belvoir, Virginia

BOYD, Morris J. (Morrie), Major General, Chief, Legislative Liaison, Office of the Secretary of the Army, Washington, DC, later Deputy Commanding General, III Corps and Fort Hood, Texas

BRAMLETT, David A., General, Commanding General, US Army Forces Command, Fort McPherson, Georgia

BROWN, Daniel J., Major General, Commanding General, US Army Combined Arms Support Command, Fort Lee, Virginia (succeeded GUEST)

BRUEN, Sheila, Captain, Commander, A Company, 16th Ordnance Battalion, 61st Ordnance Brigade, US Army Ordnance Center and Schools, Aberdeen Proving Ground, Maryland, later my Aide de Camp (succeeded STEPHENS)

CHILDRESS, Russell L. (Russ), Department of the Army Civilian, Civilian Deputy Commandant, US Army Ordnance Center and Schools, Aberdeen Proving Ground, Maryland

CLARK, Mary Joe, Major later Lieutenant Colonel, Deputy Commander, 61st Ordnance Brigade, US Army Ordnance Center and Schools, Aberdeen Proving Ground, Maryland

CLINTON, William Jefferson (Bill), President of the United States

COBURN, John G., Lieutenant General later General, Deputy Chief of Staff for Logistics, US Army, Washington, DC, later

Commanding General, US Army Materiel Command, Alexandria, Virginia (succeeded WILSON)

COHEN, William S., Presidential Appointee, Secretary of Defense, Washington, DC (succeeded PERRY)

CRAVENS, James J., Jr., Major General, Chief of Staff, US Army Training and Doctrine Command, Fort Monroe, Virginia

CROUCH, William W., General, Vice Chief of Staff, Office of the Chief of Staff, US Army, Washington, DC (succeeded GRIFFITH, replaced by SHINSEKI)

DAVID, Cecily, Colonel, Commander, Kirk Army Health Clinic, Aberdeen Proving Ground, Maryland, later Command Surgeon, US Army Forces Command, Fort McPherson, Georgia

DICKINSON, Thomas R., Brigadier General, Commanding General, 13th Corps Support Command, Fort Hood, Texas, later Commanding General, US Army Ordnance Center and Schools, Aberdeen Proving Ground, Maryland (succeeded SHADLEY)

DOHERTY, Daniel A., Brigadier General, Commanding General, US Army Criminal Investigation Command, Fort Belvoir, Virginia

DUBIA, John A., Lieutenant General, Director of the Army Staff, Washington, DC (replaced by PICKLER)

EBBESEN, Samuel E., Lieutenant General, Deputy Assistant Secretary of Defense for Military Policy, Office of the Secretary of Defense, Washington, DC

EHRLICH, Robert L. (Bob), US Representative, 2nd District of Maryland

FISHER, George A., Jr., Lieutenant General, Chief of Staff, US Army Forces Command, Fort McPherson, Georgia (replaced by PICKLER)

FOOTE, Evelyn P., Brigadier General, Special Assistant to the Secretary of the Army, Office of the Secretary of the Army, Washington, DC

FRANCE, Edward W., Jr. (Buzz), Colonel, Staff Judge Advocate, US Army Test and Evaluation Command, Aberdeen Proving Ground, Maryland

FRIFIELD, Julia, Civilian, Member of the Staff of Senator Barbara A. MIKULSKI

GAMBLE, Wayne A., Staff Sergeant later Private, Drill Sergeant, C Company, 16th Ordnance Battalion, Aberdeen Proving Ground, Maryland

GAYLORD, Robert E., Colonel, Deputy Chief of Public Affairs, Office of the Chief of Public Affairs, Office of the Secretary of the Army, Washington, DC

GIBSON, Susan S., Major later Lieutenant Colonel, Deputy Staff Judge Advocate, US Army Test and Evaluation Command, Aberdeen Proving Ground, Maryland, and my lawyer

GILL, Clair F., Major General, Commanding General, US Army Engineer Center and Fort Leonard Wood, Missouri

GLANTZ, Roslyn (Roz), Colonel, Commander, US Army Garrison, Aberdeen Proving Ground, Maryland

GLISSON, Henry T. (Tom), Major General, Commanding General/ Commandant, US Army Quartermaster Center and School, Fort Lee, Virginia (succeeded GUEST, replaced by WRIGHT)

GOOCH, Susan H., Department of the Army Civilian, Chief of Protocol, US Army Ordnance Center and Schools, Aberdeen Proving Ground, Maryland

GOODWIN, Paul, Captain, Executive Officer to the Commanding General, US Army Ordnance Center and Schools, Aberdeen Proving Ground, Maryland

GRANT, Janice, President, Harford County (MD) Chapter, National Association for the Advancement of Colored People

GRIFFIN, Natalie, Lieutenant Colonel, Office of the Staff Judge Advocate, US Army Forces Command, Fort McPherson, Georgia, and my ethics adviser

GRIFFITH, Ronald H., General, Vice Chief of Staff, Office of the Chief of Staff, US Army, Washington, DC (replaced by CROUCH)

GUEST, Robert K. (Ken), Major General, Commanding General/ Commandant, US Army Quartermaster Center and School, Fort Lee, Virginia, (replaced by GLISSON), later Commanding General, US Army Combined Arms Support Command and Fort Lee, Virginia (succeeded ROBISON, replaced by BROWN)

HALE, David R., Major General later Brigadier General, Deputy Commanding General, Allied Forces Southern Europe, later Deputy the Inspector General, Office of the Secretary of the Army, Washington, DC

HAMILTON, Mark R., Major General, Commanding General, US Army Recruiting Command, Fort Knox, Kentucky

HARKEY, William, Lieutenant Colonel, Office of the Chief of Public Affairs, Department of the Army, Washington DC

HARTSFIELD, Phil, Major, Provost Marshal, Aberdeen Proving Ground, Maryland

HARTZOG, William W., General, Commanding General, US Army Training and Doctrine Command, Fort Monroe, Virginia

HATTEN, Jim, Colonel, Staff Judge Advocate, US Army Forces Command, Fort McPherson, Georgia

HAWLEY, Thomas E., Lieutenant Colonel, Office of the Chief of Legislative Liaison, Office of the Secretary of the Army, Washington, DC

HAYDEN, Donald A. (Don), Chief Warrant Officer 3, Special Agent in Charge, Aberdeen Resident Agency, US Army Criminal Investigation Command, Aberdeen Proving Ground, Maryland

HENRY, Patrick T., Lieutenant Colonel (Retired), later Presidential Appointee, Staffer, Senate Armed Services Committee, later Assistant Secretary of the Army for Manpower and Reserve Affairs, Washington, DC (succeeded LISTER)

HOGGE, Donald J., Lieutenant Colonel, Commander, 16th Ordnance Battalion, Aberdeen Proving Ground, Maryland

HOLLOWAY, Gary A., Department of the Army Civilian, Public Affairs Officer, US Army Test and Evaluation Command and Aberdeen Proving Ground, Maryland

HOOPER, Thomas A. (Tom), Colonel, Commander, 59th Ordnance Brigade, Redstone Arsenal, Alabama (replaced by LUTTRELL)

HOSTER, Brenda L., Sergeant Major (Retired), surfaced initial allegations against Sergeant Major of the Army Gene C. MCKINNEY

HUFFMAN, Walter B., the Judge Advocate General, US Army, Washington, DC (succeeded NARDOTTI)

JACKSON, Alicia, Captain, Commander, Company C, 143rd Ordnance Battalion, Edgewood Arsenal, Maryland (succeeded ROBERTSON)

JORDAN, Larry R., Major General later Lieutenant General, Deputy the Inspector General, Office of the Secretary of the Army, Washington, DC (succeeded SIEGFRIED), later the Inspector General, Office of the Secretary of the Army, Washington, DC (succeeded BATES)

KEANE, John M. (Jack), Lieutenant General later General, Commanding General, XVIII Airborne Corps and Fort Bragg, North Carolina, later Vice Chief of Staff, Office of the Chief of Staff, US Army, Washington, DC (succeeded SHINSEKI)

KENNEDY, Claudia J., Major General later Lieutenant General, Assistant Deputy Chief of Staff for Intelligence, later Deputy Chief of Staff for Intelligence, US Army, Washington, DC

KILGALLIN, William F. (Bill), Lieutenant Colonel, Defense Counsel, US Army Trial Defense Service, Fort Gordon, Georgia, and my lawyer

KRAUER, Robert W. (Rob), Department of the Army Civilian, Deputy Provost Marshal, Aberdeen Proving Ground, Maryland

KUSSMAN, Michael J., Brigadier General, Commander, Walter Reed Army Hospital, Washington, DC

LEAVITT, Thomas P. (Tom), Colonel, Office of the Inspector General, US Army, Washington, DC

LISTER, Sara E., Presidential Appointee, Assistant Secretary of the Army for Manpower and Reserve Affairs, Washington, DC (replaced by HENRY)

LONGHOUSER, John M., Major General, Commanding General, US Army Test and Evaluation Command and Aberdeen Proving Ground, Maryland

LOPRESTI, Thomas (Tom), Colonel, Office of the Inspector General, US Army, Washington, DC

LUTTRELL, Jerry, Colonel, Deputy Commandant, US Army Ordnance Center and Schools, Aberdeen Proving Ground, Maryland (replaced by ALLEN), later Commander, 59th Ordnance Brigade, Redstone Arsenal, Alabama (succeeded HOOPER)

MCKINNEY, Gene C., Sergeant Major of the Army later Master Sergeant, Office of the Chief of Staff, US Army, Washington, DC

MCKINNEY, James C., Command Sergeant Major, US Army Training and Doctrine Command, Fort Monroe, Virginia, later Co-Acting Sergeant Major of the Army

MCLAURIN, John P., Department of the Army Civilian, Office of the Assistant Secretary of the Army for Manpower and Reserve Affairs, Washington, DC

MELTON, Clayton E. (Clay), Brigadier General, Director, Human Resources, Office of the Deputy Chief of Staff for Personnel, US Army, Washington, DC

MERRIHEW, Gerry, Command Sergeant Major, Regimental Command Sergeant Major, US Army Ordnance Center and Schools, Aberdeen Proving Ground, Maryland

MEYER, John G., Jr. (Gil), Brigadier General later Major General, Chief of Public Affairs, Office of the Secretary of the Army, Washington, DC

MFUME, Kweisi, President, National Association for the Advancement of Colored People

MIKULSKI, Barbara A., US Senator from Maryland

MILLER, John E., Lieutenant General, Deputy Commanding General, US Army Training and Doctrine Command, Fort Monroe, Virginia

MILLER, Paul David, Admiral (USN), Commander-in-Chief, US Atlantic Command, Norfolk, Virginia, later Chairman and Chief Executive Officer, Alliant Techsystems, Inc., Hopkins, Minnesota

MILLER, William, Command Sergeant Major, 61st Ordnance Brigade, US Army Ordnance Center and Schools, Aberdeen Proving Ground, Maryland

MONROE, James W. (Jim), Major General, Commanding General, US Army Ordnance Centers and Schools, Aberdeen Proving Ground, Maryland, (replaced by SHADLEY), later Commanding General, US Army Industrial Operations Command, Rock Island, Illinois

NARDOTTI, Michael J. (Mike), Major General, the Judge Advocate General, US Army, Washington, DC (replaced by HUFFMAN)

NYE, Carol, Department of the Army Civilian, my Administrative Assistant and right-hand person

PANETTA, Leon, Secretary of Defense (2011–13), Washington, DC

PICKLER, John M., Lieutenant General, Chief of Staff, US Army Forces Command, Fort McPherson, Georgia, later Director of the Army Staff, Washington, DC (succeeded DUBIA)

PORTER, John, Civilian, Member of the Staff of Senator Paul S. SARBANES

REIMER, Dennis J., General, Chief of Staff, US Army, Washington, DC (replaced by SHINSEKI)

RHAME, Thomas G., Lieutenant General, Director, Defense Security Assistance Agency, Washington, DC

RIESCO, Gabriel, Jr. (Gabe), Lieutenant Colonel, Chief of Staff, US Army Ordnance Center and Schools, Aberdeen Proving Ground, Maryland

ROBERTS, A. Reneé, Captain later Major, Staff Officer, Office of the Deputy Chief of Staff for Logistics, Department of the Army, Washington, DC

ROBERTSON, Derrick A., Captain later Private, Commander, A Company, 143rd Ordnance Battalion, Edgewood Arsenal, Maryland, (replaced by JACKSON)

ROBISON, Thomas W., Major General, Commanding General, US Army Combined Arms Support Command, Fort Lee, Virginia, (replaced by GUEST)

RODON, Raymond L., Colonel, Commander, 23rd Quartermaster Brigade, Fort Lee, Virginia

ROSS, Jimmy D., General (Retired), former boss, mentor, and friend

ROYALTY, Pamela J. (Pam), Major, Chief, Mental Health Services, Kirk Army Health Clinic, Aberdeen Proving Ground, Maryland

SARBANES, Paul S., US Senator from Maryland

SCHEMPF, Bryan H., Staff Judge Advocate, US Army Training and Doctrine Command, Fort Monroe, Virginia

SHALIKASHVILI, John M., General, Chairman of the Joint Chiefs of Staff, Washington, DC

SHINSEKI, Eric K., Lieutenant General later General, Deputy Chief of Staff for Operations and Plans, US Army, Washington, DC, later Vice Chief of Staff, US Army, Washington, DC, (succeeded CROUCH), later Chief of Staff, US Army, Washington, DC, (succeeded REIMER)

SHOLTES, J. P., Legislative Assistant to EHRLICH

SIEGFRIED, Richard S. (Steve), Major General, Commanding General, US Army Training Center and Fort Jackson, South Carolina, (replaced by BOLT), later Deputy the Inspector General, Office of the Secretary of the Army, Washington, DC, (replaced by JORDAN), later retired and then recalled to active duty to serve as Chairman, Sexual Harassment Senior Review Panel, Office of the Secretary of the Army, Washington, DC

SIMPSON, Delmar G., Staff Sergeant later Private, Drill Sergeant, A Company, 143rd Ordnance Battalion, Edgewood Arsenal, Maryland

SMITH, John A., Colonel, Office of the Chief of Public Affairs, US Army, Washington, DC

STARNES, Edward (Ed), Department of the Army Civilian, Public Affairs Officer, US Army Ordnance Center and Schools, Aberdeen Proving Ground, Maryland

STEPHENS, Jerry D., Captain later Major, my Aide de Camp (replaced by BRUEN)

STEVENSON, Mitchell H., Major General, Chief of Ordnance (2000–03), Aberdeen Proving Ground, Maryland

TAYLOR, Frederick A. (Fred), Junior Vice President, later National Commander, Military Order of the Purple Heart

TAYLOR, Tom, Department of the Army Civilian, Office of the General Counsel, Office of the Secretary of the Army, Washington, DC

THOMPSON, Linda, Lieutenant Colonel, Staff Officer, Office of the Assistant Secretary of the Army for Manpower and Reserve Affairs, Washington, DC

VICKERS, Charles, Department of the Army Civilian, Security Officer, US Army Ordnance Center and Schools, Aberdeen Proving Ground, Maryland

VOLLRATH, Frederick V., Lieutenant General, Deputy Chief of Staff for Personnel, US Army, Washington, DC

WARREN, Leroy, National Board Member (Criminal Justice Committee), National Association for the Advancement of Colored People

WEBB, Dennis M., Colonel, Commander, 61st Ordnance Brigade, US Army Ordnance Center and Schools, Aberdeen Proving Ground, Maryland

WEST, Togo G., Jr., Presidential Appointee, Secretary of the Army, later Secretary of Veterans Affairs, Washington, DC

WHITE, John P., Presidential Appointee, Deputy Secretary of Defense, Washington, DC

WILSON, Johnnie E., General, Commanding General, US Army Materiel Command, Alexandria, Virginia

WILSON, Robert (Bob), Colonel later Brigadier General, Executive Officer to the Commanding General, US Army Training and Doctrine Command, Fort Monroe, Virginia

WOOTEN, Ralph G., Major General Commanding General, US Army Chemical and Military Police Centers, Fort McClellan, Alabama

WRIGHT, James M., (Chickenman), Major General, Commanding General/Commandant, US Army Quartermaster Center and School, Fort Lee, Virginia

ABBREVIATIONS
AND ACRONYMS

1SG: First Sergeant (E-8)

AA: Appointing Authority

ABCMR: Army Board for Correction of Military Records

AIT: Advanced Individual Training

AMC: Army Materiel Command

APG: Aberdeen Proving Ground

AR: Army Regulation

ASA(M&RA): Assistant Secretary of the Army for Manpower and Reserve Affairs

ATK: Alliant Techsystems, Inc.

AUSA: Association of the US Army

AWOL: Absent Without Leave

Bde: Brigade

BG: Brigadier General (0-7, One-Star)

Bn: Battalion

CASA: Civilian Aide to the Secretary of the Army

CASCOM: Combined Arms Support Command

CAT: Crisis Action Team

CG: Commanding General

CID: Criminal Investigation Command

CJCS: Chairman of the Joint Chiefs of Staff

CODEL: Congressional Delegation

COL: Colonel (O-6)

CPT: Captain (0-3)

CSA: Chief of Staff of the Army

CSM: Command Sergeant Major (E-9)

CW3: Chief Warrant Officer Three

CW5: Chief Warrant Officer Five

DA: Department of the Army

DAIG: (Office of the)Department of the Army Inspector General

DC: Division Commander

DCG : Deputy Commanding General

DCSOPS: Deputy Chief of Staff for Operations (G-3)

DEOMI: Defense Equal Opportunity Management Institute

DOD: Department of Defense

DS: Drill Sergeant

DTIG: Deputy (to) the Inspector General

DTLOMS: Doctrine, Training, Leader Development, Organization, Materiel, and Soldiers

DWI: Driving While Intoxicated

EA: Edgewood Arsenal (a sub-post of APG)

EO: Equal Opportunity

EOD: Explosive Ordnance Disposal

FORSCOM: Forces Command

GAO: General Accountability Office

GCM: General Court-Martial

GCMCA: General Court-Martial Convening Authority

GEN: General (O-10, Four Star)

GO: General Officer

GOMOR: General Officer Memorandum of Reprimand

HNSC: House National Security Committee

IET: Initial Entry Training

IG: Inspector General

IO: Investigating Officer

KGB: National Security Agency of the Soviet Union 1954–1991

LogSME: Logistics Subject Matter Expert

LOR/MOR: Letter of Reprimand

LTC: Lieutenant Colonel (O-5)

LTG: Lieutenant General (O-9, Three-Star)

MACOM: Major Command (of the Army)

MAJ: Major (O-4)

MEB: Medical Evaluation Board

MEOCS: Military Equal Opportunity Climate Survey

MG: Major General (O-8, Two-Star)

MOR/LOR: Memorandum of Reprimand

MOS: Military Occupational Specialty

MP: Military Police

NAACP: National Association for the Advancement of Colored People

NCO: Noncommissioned Officer (E-5 through E-9)

NTC: National Training Center

OCLL: Office of the Chief of Legislative Liaison

Ord: Ordnance

OMPF: Official Military Personnel File

OSD: Office of the Secretary of Defense

PAO: Public Affairs Office/Public Affairs Officer

PVT/Pvt: Private (PV1 = E-1 and PV2 = E-2)

Q&A: Question and Answer

R&A: Review and Analysis

RSA: Redstone Arsenal, Alabama

SA: Secretary of the Army

SASC: Senate Armed Services Committee

SECDEF: Secretary of Defense

SFC: Sergeant First Class (E-7)

SGM: Sergeant Major (E-9)

SGT: Sergeant (E-5)

SJA: Staff Judge Advocate

SMA: Sergeant Major of the Army (E-9)

SSG: Staff Sergeant (E-6)

TC: TRADOC Commandant

TIG: The Inspector General of Office of the Department of the Army Inspector General

TJAG: The Judge Advocate General of the Army

TRADOC: Training and Doctrine Command

UCMJ: Uniform Code of Military Justice

USAF: US Air Force

USAOC&S: US Army Ordnance Center and School

USAOMMC&S: US Army Ordnance Missile and Munitions Center and School

USMC: US Marine Corps

VCSA: Vice Chief of Staff of the Army

Wandas: Women Active in Our Nation's Defense

TERMS, DEFINITIONS, AND REFERENCES

ABERDEEN PROVING GROUND

In 1995–97, Aberdeen Proving Ground (APG) was a sprawling military base of 72,516 acres along the Chesapeake Bay and adjacent to the city of Aberdeen in Harford County, Maryland. The sexual misconduct that occurred at the US Army Ordnance Center and School (USAOC&S), only one of many tenant units on the base, is sometimes called the "Army's Aberdeen Sex Scandal." But that really does a disservice to the other units and their personnel on the APG installation and to the fine citizens of the city of Aberdeen, which was named an All American City in 1997. Edgewood Arsenal was considered a sub-post of APG. Of the 11,692 military and civilian employees working on APG on a daily basis, 3,317 were in the USAOC&S. (Source: Facts and Figures: Aberdeen Proving Ground, Maryland, October 1, 1996, Headquarters, Aberdeen Proving Ground, Maryland, ATTN: STEAP-RM-MP)

ARTICLE 15

Nonjudicial punishment under Article 15 of the Uniform Code of Military Justice (UCMJ) is an administrative tool used to maintain discipline. It is imposed by a commander on a soldier for offenses committed under the UCMJ. There are three levels of Article 15:

Summarized Article 15: This is the least serious of the three and can be used only on enlisted soldiers. Maximum punishment, if found guilty at summarized proceedings, is 14 days of restriction and 14 days of extra duty. Forfeiture of pay and reduction in rank cannot be adjudicated during summarized proceedings. Summarized Article 15s cannot be filed in a soldier's official military personnel file (OMPF).

Company Grade Article 15: Maximum punishment is 14 days of extra duty, 14 days of restriction, and forfeiture of 7 days of pay. Soldiers in the grade of E-4 (specialist 4) and below can be reduced one pay grade. Punishments may be suspended for up to six months (suspension is used to grant a probation period). The record of proceedings may be filed in the soldier's OMPF.

Field Grade Article 15: These proceedings are conducted by a commander in the grade of major or above (that is, a field grade officer). Maximum punishment is 60 days of restriction (unless imposed in conjunction with extra duty, then the maximum is 45 days of restriction), 45 days of extra duty, and forfeiture of ½ of one month's pay for two months. Soldiers in the grade of E-4 and below may be reduced one or more grades; soldiers in the grade of E-5 (sergeant) or E-6 (staff sergeant) may be reduced one grade. Punishments may be suspended for up to six months. The record of proceedings may be filed in a soldier's OMPF.

For all Article 15s, a soldier has the right to refuse to have the case heard at an Article 15 proceeding and demand a trial by court-martial. For company grade and field grade Article 15s, soldiers may consult with defense counsel before making a decision about whether to accept proceedings under Article 15. Defense counsel will also offer advice to the soldier on how to present the case if they both agree to have the case heard at the Article 15 proceeding.

At an Article 15 hearing, the soldier has the right to remain silent, present evidence, call witnesses, be accompanied by a spokesperson, request an open hearing, and examine the available evidence.

If a soldier is found guilty at the Article 15 hearing, the soldier has the right to appeal the findings and punishment to the next higher commander. (Source: Major Susan S. Gibson, February 19, 1997)

ARTICLE 32 INVESTIGATION

While the Article 32 UCMJ investigation is often called the military's counterpart to the civilian grand jury, it provides the accused with procedural rights not found in grand jury proceedings. For example, testimony, documents, and other evidence considered by the investigating officer are available to the defense, while the proceedings in a civilian grand jury are usually off-limits. The accused also is afforded counsel and attends the investigation. The accused and counsel do not have this right in civilian practice. In addition, the defense may cross-examine all witnesses. An impartial investigation is a substantial right. (Source: APG News Release No. 5, February 1997)

CHAPTER DISCHARGES

The expression "chaptered out of the Army" is military shorthand talk for the individual being separated from active duty under the provisions of one of the chapters in Army Regulation 635-200, "Active Duty Enlisted Administrative Separation." The most common chapters, in addition to Chapter 10 (see below), in 1995–97 were:

Chapter	Title
5	Separation for the Convenience of the Government
6	Because of Dependency or Hardship
8	Separation of Enlisted Women—Pregnancy
9	Alcohol or Other Drug Abuse Rehabilitative Failure
11	Entry Level Performance and Conduct
13	Separation for Unsatisfactory Performance
14	Separation for Misconduct
15	Discharge for Homosexual Conduct
18	Failure to Meet Body Fat Standard
19	Qualitative Management Program

CHAPTER 10: DISCHARGE IN LIEU OF COURT-MARTIAL

This type of discharge is governed by Chapter 10 of the Army Enlisted Personnel Regulation (AR 635-200). Any soldier who is facing UCMJ charges for an offense that can be punished by a bad conduct discharge or a dishonorable discharge may request a Chapter 10 discharge in lieu of a court-martial (also called a discharge for the good of the service).

When requesting a Chapter 10 discharge, the soldier must sign a document stating that he understands that he may be discharged under other than honorable conditions, that he may be deprived of many or all Army and Veterans Administration benefits, and that this may cause substantial prejudice in civilian life.

The regulation states that a "discharge under other than honorable conditions" normally is appropriate for a soldier who is discharged under Chapter 10. However, in some cases, a soldier may be granted a general discharge. It is extremely rare for a soldier to receive an honorable discharge under this chapter. A soldier who is discharged with an other than honorable discharge under Chapter 10 is also reduced to the rank of private (E-1). (Source: Major Susan S. Gibson, February 19, 1997)

DTLOMS

The primary mission of TRADOC Service School CG/Commandant/Branch chief is to oversee the development, coordination, and synchronization of their Branch DTLOMS:

- Doctrine Development
- Training Development and Execution
- Leader Development
- Organization (The Ordnance Branch chief is responsible for the tables of organization and equipment [TOEs] required for supporting the Field Army.)

- Materiel (These are systems fielded or to be fielded that require maintenance or repair training for mechanics/ repairers.)
- Soldier Support

HISTORY OF THE ORDNANCE CORPS

"The Ordnance Corps dates to the early days of the American Revolution. In 1775 a Continental Congress committee which included George Washington convened to study the methods of arms and ammunition procurement and storage. As a result, Ezekial Cheever was appointed as the Commissary General of the Artillery Stores, making him essentially the first Chief of Ordnance. In 1776, a Board of War and Ordnance was created, with the responsibility of issuing supplies to troops in the field. The next year, the first Ordnance powder magazine was established at Carlisle, Pennsylvania, followed shortly thereafter by the first arsenal and armory operations at Springfield, Massachusetts. Other arsenals and armories were also established at Harpers Ferry, Philadelphia, and Watertown, near Boston.

"On May 14, 1812, the Ordnance Department was formally organized by Congress as part of the preparations for the second British war. The department assumed responsibility for arms and ammunition production, acquisition, distribution, and storage in a much broader geographical base than in the War of Independence.

"During the War Between the States, the Ordnance Corps was seriously tested since its installations were primarily targets for operations by both sides. True to its traditions, the Corps successfully brought about massive procurement of weapons and supplies, effectively providing field support for fast moving armies.

"In the war with Spain in 1898, the Ordnance Department first deployed material overseas and provided close combat support.

"During World War I, the Ordnance Department mobilized an immense industrial base, developed weapons systems in cooperation with the allies, organized a variety of Ordnance training facilities, and established large overseas supply depots. World War II saw an even more dramatic expansion of the Ordnance mission of production, procurement, maintenance, and training. In both Korea and Vietnam, the Ordnance Corps provided materiel supply and maintenance, characteristic of its tradition of 'service to [the] line, on the line, on time' and was active in the development of rockets, guided missiles, and satellites. The operations in Grenada and Panama, as well as Operation Desert Shield/ Desert Storm showed the world that the U. S. Army is ready for any contingency. Ordnance Corps soldiers played an extensive role by providing support at all levels on the battlefield.

"With the advent of the modern U. S. Army Regimental system, the Ordnance Corps is organized under the whole branch concept. The Chief of Ordnance serves as the Regimental Commander." (Source: "U.S. Army Ordnance Corps" undated pamphlet (approx. 1996), Office Chief of Ordnance, pp. 13–14)

INITIAL ENTRY TRAINING

When Ordnance soldiers entered the Army in 1995, they underwent initial entry training (IET), which consisted of two main parts: (1) basic combat training (BCT); and (2) advanced individual training (AIT).

BCT (8 weeks in duration) was conducted at Fort Knox, Kentucky; Fort Leonard Wood, Missouri; and Fort Jackson, South Carolina.

AIT (course length for Ordnance soldiers varied with each military occupational specialty and could last anywhere from a few weeks to almost a whole year) was conducted at branch proponent schools (e.g., USAOC&S and US Army Ordnance Missile and Munitions Center and School).

Control over the trainees in IET was gradually lessened until the environment in which the trainees lived and worked in AIT equated to what it would be like in an operational unit or organization.

The drill sergeants (DSs) in IET were the noncommissioned officers (NCOs) who provided the control over the trainees. Other NCOs, warrant officers, officers, and civilians provided the technical training in

AIT. These two groups of individuals, with their leaders, were referred to as the cadre.

One of the first things a new recruit is taught in basic training is to follow the chain of command. That means, if you have a problem or concern or need help in any way, you talk first to your immediate supervisor. As a general rule, a recruit doesn't go over his or her supervisor's head to talk with someone higher up in the chain of command. In the case of soldiers in BCT and AIT in 1995–97, the first-line supervisor for the trainee was the DS.

MEMORANDUM OF REPRIMAND PROCESS

I use the terms "letter of reprimand" (LOR) and "memorandum of reprimand" interchangeably. The official title is memorandum of reprimand (MOR). An MOR signed by a general officer is a general officer memorandum of reprimand (GOMOR).

MORs are given as punishment for wrongdoing that falls below the level justifying a court-martial, but can be a punishment meted out in a court-martial (general, special, or summary) or in the Article 15 process.

A MOR can be given without going through either the court-martial or Article 15 process. The person administering the MOR prepares the document and sends it to the alleged offender, who is given some time to respond.

Based on the offender's response, the administrating official can do the following to the original MOR:

- Keep it as it is and issue it to the offender officially.
- Modify and issue it to the offender officially.
- Deem it not necessary, destroy it, and consider the matter closed.
- Deem it not necessary, destroy it, and give the offender a verbal reprimand.

When an MOR is officially given to an offender, the last paragraph provides the disposition instructions, which are basically two: (1) direct the MOR to be filed in the offender's OMPF maintained at the Department of the Army; or (2) direct the MOR to be filed at the local level.

A MOR in one's OMPF is negative information seen by promotion selection boards and, in some cases, by the Senate who confirms promotions and appointments.

MORs can be overcome, but usually not if they are highly visible.

ABOUT THE AUTHOR

Robert D. Shadley is a retired US Army major general and author of *The GAMe: Unraveling a Military Sex Scandal.*

During his leadership career, he guided more than 3,500 military men and women in combat and over 20,000 students in training in peacetime.

He retired from active duty in 2000, following a distinguished 33-year military career serving in key command and staff assignments, to include combat tours in Viet Nam and OPERATION DESERT SHIELD/STORM.

Major General Shadley then served in key leadership positions at Alliant Techsystems, Inc. (ATK). He also has served as a senior mentor providing logistics and leadership subject matter expertise to Army units prior to deployment to Afghanistan and Iraq, in addition to training exercises. He currently provides acquisition and logistics consulting advice to businesses in the aerospace and defense sector.

His numerous awards and decorations include the Distinguished Service Medal, the Defense Superior Service Medal, the Legion of Merit with two Oak Leaf Clusters, and the Bronze Star Medal with Oak Leaf Cluster.

Major General Shadley is active in community and non-profit organizations in the Wayzata, Minnesota area, where he and his wife, Ellie, reside.

To learn more about the author please visit:
www.ShadleyEditions.com